Green Supply Chain Management

Today, one of the top priorities of an organization's modern corporate strategy is to portray itself as socially responsible and environmentally sustainable. As a focal point of sustainability initiatives, green supply chain management has emerged as a key strategy that can provide competitive advantages with significant parallel gains for company profitability. In designing a green supply chain, the intent is the adoption of comprehensive and cross-business sustainability principles, from the product conception stage to the end-of-life stage. In this context, green initiatives relate to tangible and intangible corporate benefits. Sustainability reports from numerous companies reveal that greening their supply chains has helped reduce operating cost, thus boosting effectiveness and efficiency while increasing sustainability of the business.

Green Supply Chain Management provides a strategic overview of sustainable supply chain management, shedding light on the theoretical background and key principles of the topic. Specifically, this book covers various thematic areas including benefits and impact of green supply chain management; enablers and barriers on supply chain operations; inbound and outbound logistics considerations; and production, packaging and reverse logistics under the notion of "greening". The ultimate aim of this textbook is to highlight the challenges in the implementation of green supply chain management in modern companies and to provide a roadmap for decision-making in real-life cases.

Combining chapter summaries and discussion questions, this book provides an accessible and student-friendly introduction to green supply change management and will be of great interest to students, scholars and practitioners in the fields of sustainable business and supply chain management.

Charisios Achillas is a Professor at the Technical Educational Institute of Central Macedonia, Department of Logistics. His work primarily focuses on the areas of environmental engineering, reverse logistics and sustainable development.

Dionysis D. Bochtis works in the area of Systems Engineering focused on bio-production, and related provision systems, with enhanced ICT and automation technologies up to fully robotized systems. He is a Professor in Agri-Robotics at the University of Lincoln, UK, Institute for Agri-Food Technologies (LIAT).

Dimitrios Aidonis is a Professor in the Department of Logistics at the Technological Educational Institute of Central Macedonia, Greece. His research interests focus on the area of Operations Management and his work deals mainly with Supply Chain, Reverse Logistics Management and Green Logistics.

Dimitris Folinas is a Professor in the Department of Logistics at the Technological Educational Institute of Central Macedonia, Greece. His research interests and working experiences refer to logistics and supply chain management technologies and information systems.

"The authors efficiently capture all the different aspects and latest trends in green supply chain management in a condensed, practical and reader-friendly way. A must-read guide for academics, students and entrepreneurs!"

Simon Pearson, Professor and Director of LIAT at Lincoln Institute for Agri-Food Technologies, UK

"In recent years, green supply chain management has become the focus among practitioners and researchers across the globe. This work provides an excellent synopsis of the current status, as well as the upcoming trends in the field. The authors clearly explain the different angles of green supply chain management and offer practical advice towards increasing the effectiveness and adaptability of supply chains in the modern business world."

Kyriakos Kouveliotis, Professor at International Telematic University UNINETTUNO, Italy

"One of the most critical issues in respect to competitiveness of enterprises in the modern business environment is balancing economic profits with environmental performance. The pressure from the community and the customers, the constantly increasing regulatory legislation, and the profit potential from energy savings or decreased waste management costs, put green supply chain management in the forefront of business efficiency. Connecting theory with practice, the authors offer a comprehensive overview of recent developments in the thematic area of green supply chain management."

Marinella Christoforou, Managing Director at TEAM CERT Certification & Inspection Services, Greece

"This book is definitely a great work full of important information on sustainability in supply chains, which is a top agenda issue worldwide. It is an easy-to-read book, useful for the business world and academia."

Claus Aage Grøn Sørensen, Professor in the Department of Engineering at Aarhus University, Denmark

Green Supply Chain Management

Charisios Achillas,
Dionysis D. Bochtis,
Dimitrios Aidonis and
Dimitris Folinas

Routledge
Taylor & Francis Group
LONDON AND NEW YORK

earthscan
from Routledge

First published 2019
by Routledge
2 Park Square, Milton Park, Abingdon, Oxon OX14 4RN

and by Routledge
711 Third Avenue, New York, NY 10017

Routledge is an imprint of the Taylor & Francis Group, an informa business

British Library Cataloguing-in-Publication Data
A catalogue record for this book is available from the British Library

Library of Congress Cataloging-in-Publication Data
Names: Achillas, Charisios, author.
Title: Green supply chain management / Charisios Achillas, Dionysis D. Bochtis, Dimitris Aidonis, and Dimitris Folinas.
Description: Abingdon, Oxon ; New York, NY : Routledge, 2019. | Includes bibliographical references and index.
Identifiers: LCCN 2018030646 | ISBN 9781138644601 (hbk) | ISBN 9781138644618 (pbk) | ISBN 9781315628691 (ebk) | ISBN 9781317241782 (mobipocket unencrypted) | ISBN 9781317241805 (adobe reader) | ISBN 9781317241799 (epub)
Subjects: LCSH: Business logistics—Environmental aspects. | Industrial management—Environmental aspects. | Industrial ecology.
Classification: LCC HD38.5 .A297 2019 | DDC 658.7—dc23
LC record available at https://lccn.loc.gov/2018030646

ISBN: 978-1-138-64460-1 (hbk)
ISBN: 978-1-138-64461-8 (pbk)
ISBN: 978-1-315-62869-1 (ebk)

Typeset in Sabon
by Apex CoVantage, LLC

Contents

List of figures viii
List of tables x
List of acronyms xi

1 Green supply chain framework 1

2 From traditional supply chain to green supply chain 5

3 Enablers and barriers of the green supply chain 14

4 Inbound logistics (green procurement) 38

5 Green production 64

6 Outbound logistics (green transportation and distribution) 74

7 Green packaging 87

8 Reverse logistics 97

9 Decision-making in the green supply chain 136

10 Technologies of marketing and supply chain management synergies 146

References 169
Index 174

Figures

1.1	Green supply chain framework	2
2.1	Evolution of supply chains	7
2.2	Top-down, bottom-up and hybrid transformation	9
2.3	Corporate social responsibility elements	10
3.1	Internal barriers summary	16
3.2	External barriers overview	23
3.3	Internal motivators and drivers overview	27
3.4	External motivators and drivers overview	31
4.1	Benefits from green practices in procurement	43
4.2	Green procurement life cycle	45
5.1	Stages of green production	67
5.2	The ten principles of eco-design	71
6.1	Allocation of energy consumption	75
6.2	Allocation of GHG emissions	76
6.3	Green practices in transportation/distribution	83
7.1	Benefits and disadvantages of environmental labels	90
7.2	Environmental labeling process	94
7.3	Environmental labeling applications	95
8.1	Hierarchy of waste management alternative strategies	98
8.2	Product's life cycle	98
8.3	Typical composition of WEEE	101
8.4	Stream of EEE product at the end of its life cycle	104
8.5	Stages of material recovery from WEEE	105
8.6	Stages for WEEE material recovery process	106
8.7	Methodological framework for the integrated assessment of WEEE	110
8.8	Key considerations within the sustainable development concept	112
8.9	Application of Pareto law in WEEE management	113
8.10	Main sources of ECDW	116
8.11	Key issues that influence the type and quantity of ECDW	117
8.12	Ranking of ECDW management policies	120
8.13	Key technological issues in ECDW management	121

8.14 Material and information flow in the construction industry supply chain 124
8.15 Parameters that need to be considered in the design of a building 126
9.1 Assessment frameworks and methods 137
10.1 Aspects of GrICT and GSCM relations 146
10.2 Aspects of GrICT in GSCM 147
10.3 Affected areas of GrICT in GSCM 149
10.4 GrICT solutions in the market 150
10.5 Stages in the life cycle of products and services, from raw material acquisition through end-of-life disposition (cradle-to-grave) 151
10.6 RFID functional elements 156
10.7 Sensors inside the container 157
10.8 Sensors/embedded systems in “smart” container 158
10.9 ecoCycle starting page 162
10.10 Entry of data relating to the production process 163
10.11 Data entry relating to energy consumption 165
10.12 Data entry relating to waste management 166
10.13 Results of calculations 167

Tables

4.1 Best practices for the application of green procurement techniques in enterprises 42
4.2 Best practices of the application of green procurement techniques in the public sector 42
4.3 Example of scoring model 53
4.4 Weighing qualitative criteria in pairs 56
4.5 Calculation of criteria weightings 56
4.6 Assessing alternatives by criterion A 57
4.7 Assessing alternatives by criterion B 57
4.8 Assessing alternatives by criterion C 57
4.9 Assessing alternatives by criterion D 57
4.10 Assessing alternatives by criterion E 58
4.11 Overall assessment of alternatives based on qualitative criteria 59
4.12 Rating of alternatives based on quantitative criteria 59
4.13 Final assessment of alternatives 61
5.1 Simplified life cycle assessment methods 69
6.1 Basic types of environmental indicators and benefits in the context of local food supply chains 79
8.1 Hazardous materials in EEE products and disposal methods 103
8.2 Construction waste management options 122
8.3 Comparative presentation of the advantages of carrying out selective deconstruction works and conventional building demolition 133
8.4 Major restrictions on the application of selective deconstruction process 134

Acronyms

ANP	analytic network process
BSI	British Standards Institution
B2B	business-to-business
CBA	cost-benefit analysis
CDW	construction and demolition waste
CSR	corporate social responsibility
CVM	contingent valuation method
DTD	door to door
DW	data warehousing
ECDW	excavation, construction and demolition waste management
ECOEFA	eco-efficiency analysis
EEE	electrical and electronic equipment
EF	ecological footprint
EIA	environmental impact analysis
EMAS	environmental management and auditing scheme
EMS	environmental management system
EPD	environmental product declaration
ESPD	European single procurement document
GHG	greenhouse gases
GSC	green supply chain
GSCM	green supply chain management
HPM	hedonic pricing method
ICT	information and communication technologies
IoT	Internet of Things
IRR	internal rate of return
ISO	International Organization for Standardization
ITS	intelligent transportation systems
JIT	just in time
KEPI	key environmental performance indicator
LCA	life cycle analysis
LCC	life cycle costing
LCSD	life cycle sustainability dashboard

MCA	multi-criteria analysis
MIPS	material intensity per service unit
NPV	net present value
OLAP	online analytical processing
QASI	quantitative assessment of sustainability indices
RFI	request for information
RFID	radio frequency identification
RFP	request for proposal
RFQ	request for quote
R&D	research and development
SEA	strategic environmental assessments
SME	small and medium enterprise
SPP	sustainable public procurement
SRG	sustainability reporting guidelines
TCO	total cost of ownership
TQM	total quality management
VMI	vendor-managed inventory
VOC	volatile organic compounds
WEEE	waste electrical and electronic equipment
3PL	third-party logistics

1 Green supply chain framework

Discussion questions

- Why is a multidisciplinary approach critical for green supply chain management?
- Which are the key dimensions and disciplines of green supply chains?
- How are different research topics in green supply chain management interlinked within an integrated framework?

Green supply chain (GSC) includes policies, practices and tools that an organization can apply in the context of the sustainable environment. Even it being the case that the integration of environmental concerns within supply chain management has itself evolved into a separate research and business field, GSC can be considered as an interdisciplinary topic, involving different and multiple objectives of business, social, economic, technological and environmental sustainability issues. To assist the advancement of the multidisciplinary research field of GSC, a framework is provided in this first chapter to understand and appreciate the relationships of various research topics in this field. As numerous aspects of supply chain activities are examined, inevitably the multidisciplinary nature of the system will emerge. Each facet of the system is served by a combination of disciplines that come into focus as the particular subsystem is delineated. The proposed framework also acts as a roadmap for the chapters of the book, aiming to act as an integrated prism for the various research disciplines.

Appreciate the relationships of various research topics in the green supply chain

As a focal part of sustainability initiatives, green supply chain management (GSCM) has emerged as a key strategy that can provide competitive advantage with significant gains for the company's bottom line.[1] In designing green supply chains, the intent is to adopt best practices comprehensively and across business boundaries, from product conception to the end-of-life recycling stage. In this context, green initiatives relate to tangible and intangible corporate benefits. Sustainability reports of many companies indicate that the greening of their supply chains has assisted them in reducing their operating costs, with increased sustainability of their business.

Greater importance of inter-organizational relationships has caused organizations to consider building competitive advantage via management of their supplier and customer partnerships and networks. This evolution in management and business focus resulted in development of the supply chain and supply management fields. In this introductory chapter, a conceptual framework and theoretical background is presented. Utilizing this framework, emergent research directions to advance the field are also presented. The structure of the textbook will be based on the proposed framework.

The integrated planning of the green supply chain requires the management of a business or organization to initially determine the inputs, drivers and enablers that must be processed for the production, transportation and distribution, packaging and recycling of green products (Figure 1.1).

The management of green supply includes the planning, execution, monitoring and control of practices, approaches and tools that assists organizations of their "greening" process to become socially responsible and sustainable through environmental protection.

Another critical issue is the identification of the key stakeholders within GSC initiatives. Sustainable supply chain management expands the concept of sustainability from a company to the supply chain level by providing companies with tools for improving their own and the sector's competitiveness, sustainability and responsibility towards meeting stakeholder expectations. Principles of accountability, transparency and stakeholder engagement are highly relevant to sustainable supply chain management. During recent decades, a number of innovative practices and technologies have emerged

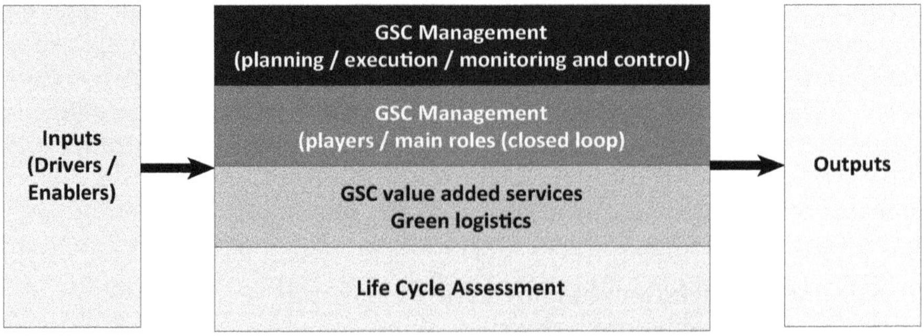

Figure 1.1 Green supply chain framework

to achieve the automation, simplification, optimization and redesign of GSCM processes. Specifically, the following initiatives have been promoted: (a) procurement-sourcing, manufacturing, re-manufacturing, warehousing, supply chain network design and waste management; (b) improving the communication and achieving the coordination, cooperation and integration of the supply chain partners of the supply chain; and (c) supporting the decision-making process in the three business levels (operational, tactical and strategic). Moreover, there is a need to identify the outputs and/or services, but also the social, financial and environmental benefits.

This conceptual framework acts as a roadmap for the topics of the book, aiming to provide an integrated prism instead of a self-directive and isolated study of the aforementioned key areas.

Green supply chain book roadmap

Green supply chain focuses not only on cost, efficiency and high customer service, but also on low environmental consequences. Chapter 2 provides the unique characteristics of green supply chain and discusses the main stages of the evolution from the traditional supply chain to the green supply chain. The identification and classification of the drivers and enablers to green and add sustainability in the supply chain, as well as the understanding of their mutual relationships, are the main objectives of Chapter 3.

In Chapter 4, the functional area of procurement in the context of green supply chain is examined. The chapter aims to define green procurement, as well as to identify the economic and environmental concerns that have contributed to increasing interest in green procurement. One of the key topics of the chapter is the description of the green procurement life cycle (green procurement cycle stages). Furthermore, a number of cases of green procurement initiatives and the barriers to its broader adoption are presented as examples.

Green production is the examined topic in Chapter 5. In this chapter, green production (manufacturing) processes and the corresponding production cycle are discussed. Sustainable materials, modern production techniques, technologies and applications are also presented, aiming first to assess the environmental impact of materials, manufacturing processes and product life cycles, and second to sketch the green production portfolio within a focused factory.

Chapter 6 describes the third functional area: transportation and distribution. Priorities and objectives of sustainable transportation, as well as policies, best practices and technologies for greening the transportation and distribution processes are presented and analyzed.

Green (sustainable) packaging is examined in Chapter 7. Packaging materials, policies and regulations are discussed. Case studies and new technologies, materials and processes are identified in order to better describe the eco-friendly packaging procedure.

Waste reduction is a critical success factor in green supply chain management, and specifically in the reverse logistics. In this context, the concepts of reverse logistics and closed loop are defined. Waste reduction strategies, best practices and example cases are also presented in Chapter 8.

Planning using sustainability criteria and multiple-criteria (cost-environment) planning in the three business levels is examined in Chapter 9. Drivers behind green supply chain strategies, as well as barriers and motivators are also presented. Moreover, performance measurement methods for green supply chain initiatives are discussed and key environmental performance indicators (KEPIs) are identified.

In Chapter 10, future trends, challenges and issues influencing green business decisions are presented. Moreover, the role of information and communication technologies (ICT) in the transition of a conventional supply chain to a green supply chain is estimated. Green information and communication technologies (GrICT), a term which refers to all the technological solutions that can be used to improve environmental performance throughout the economy and society, is presented. Case studies and best practices, as well as management of green technologies, are also discussed.

Chapter summary

In this chapter, we have provided a framework for green supply chain management, and presented the relationships between various research themes that are closely interlinked within the topic.

2 From traditional supply chain to green supply chain

Discussion questions

- How have supply chains evolved over the years?
- What motivated the need for improvements in supply chain management?
- How are green supply chains differentiated from conventional ones?
- What are the different types of supply chains that have emerged over the years?
- Why is environmental sustainability being integrated in modern business strategy?
- Why do supply chains at a global scale introduce corporate social responsibility in their business strategies?
- Why does transforming a conventional supply chain into a green supply chain not always result in environmental savings? What are the paradoxes in green supply chains?

Supply chains in ancient times

In ancient times, manufacturing of goods was directly dependent on the accessibility of raw materials, while their moving was obstructed by limitations in transportation technology. As a result, most of the products were produced and consumed at a local level, without major transportation requirements. Also, since transportation of goods was an expensive process, it was limited to the most valuable goods, such as gold and weapons. Supply chains were simple, restricted mainly to the immediate interactions between producers and consumers.

However, the concepts of logistics and supply chain management were being applied even from the ancient times. A representative example is the one of Alexander the Great, whose impressive achievements were not only based on his army's competence, but above all on the application of primary – but very efficient – management of his supply chain. For instance, Alexander's the Great army camped near rivers or seaports in order to facilitate the transportation of substantial amounts of supplies from other areas of his empire. According to the historian Donald Engels,[2] Alexander the Great planned his moves and managed his army taking advantage of the strong parts of his army's supply chain.

In the last 100 years, supply chains have been evolving in order to respond to current, ever-changing market needs. In the material to follow within this chapter, we aim to present the evolution of traditional supply chains and their management from its origins to the most recent transition to green supply chains. We also attempt to explain the motivators of this transition.

The origins of supply chains

As mentioned earlier, supply chains have been applied even from the ancient years. In the last century, supply chains have been developed and adjusted to meet the global market needs. More specifically, in the beginning of the 20th century, improvements in supply chain management aimed at facilitating processes that required a large amount of workforce. Today, supply chains can be extremely complicated, involving an immense number of stakeholders with different expectations and needs, from raw material manufacturers to end customers. Thus, modern supply chains require a more strategic and integrated management for their efficient operation.

Supply chain management attempts to assess all the components of a supply chain, such as information and material flows. The need for supply chain management triggered the birth and evolution of both industrial engineering and operations research. During the 1910s, Fredrick Taylor, who is considered the father of industrial engineering, aimed his study on the improvement of hand-operated loading procedures. The complex needs of army supply chain management in the Second World War during the 1940s led to the development of operation research and the recognition of analytics as a valuable tool for the relevant studies. When put together, industrial engineering and operation research can provide effective solutions regarding supply chain and logistics issues.

The evolution of supply chains

In a conventional supply chain, a product's journey starts from the raw material supplier. Then the manufacturer converts the materials into finished or semi-finished products, which are then brought to the distributors or the wholesalers. Then, based on the market demand, products are sold to the retailers, and finally they end up with the final customers/consumers. The previously mentioned procedures, with the involvement of all relevant stakeholders (i.e. suppliers, manufacturers, forwarders, retailers, customers, etc.) constitute a supply chain.

Over the years, market expansion and globalization led to the consequent evolution of supply chains and their management. Companies need to respond to a variety of challenges, especially in a constantly changing global market. Over recent decades, international trade has evolved rapidly. Customers get more and more knowledge about available products and services, and increasingly require high quality at lower costs. In addition, the evolution of retail marketing (e.g. 24-hour shops, personal deliveries, e-shops, etc.) creates the demand of more efficient supply chain management in order to satisfy consumer needs.

Businesses need to maintain their competitiveness since the competition is getting relentless. To that end, enterprises turn to the development of their entire supply chains and not only the products themselves. Competition is now between supply chains, rather than between enterprises. One of the main purposes for the establishment of a supply chain is to increase its stakeholder's revenue by minimizing the total cost, as well as the response time for the consumers. Small response times is one of the main customer demands in recent years. It should be also highlighted that integration of organizations has emerged to be an important motivator for the increase of an organization's competitiveness.

The evolution of supply chains over the years was highly dependent on technological evolution and customer needs, as well as on the current market requirements. In Figure 2.1, several types of supply chains that have been developed over the years are illustrated.

The first type of supply chain, the **standard supply chain**, was designed to meet the customers' needs for standard products that did not require

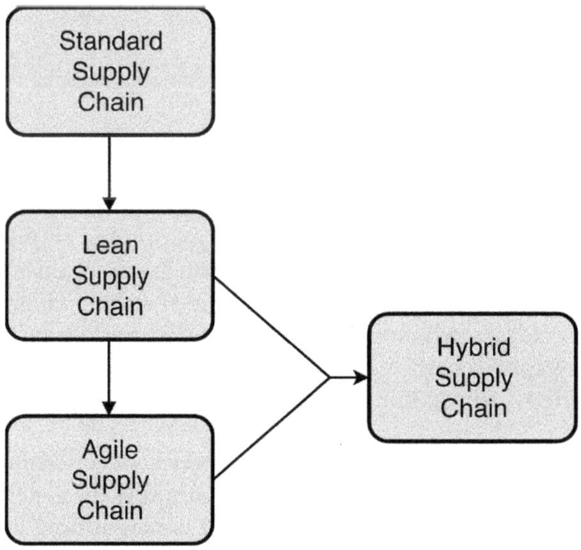

Figure 2.1 Evolution of supply chains

adaptability and resource preservation. Standardized products are products that have fixed demand and do not have alterations in their traits or their design. The procurement of raw materials for their production is realized in copious quantities, which created the need for standard suppliers.

Later in history, the **lean supply chain** was developed as an improvement on the standard supply chain. The lean supply chain is aiming at ongoing development, mitigation of waste and generally at eradicating all the actions that do not create added value for the consumers. Lean supply chains consider all the life cycle stages of a standard product and attempt to increase the supply chain's efficiency by decreasing the number of steps in the procedure, the time needed and also, all the relevant costs and waste. Also, this type of supply chain can generally satisfy more specialized customer needs and support the production of products in small quantities. The major disadvantage of a lean supply chain is that it is not adaptable to unpredictable situations.

In order to address the deficiencies of these two supply chains, the **agile supply chain** was developed. An agile supply chain was initially designed to be flexible in unpredictable situations and adaptable to the respective market needs. In addition, agile supply chains can rapidly respond to the unexpected demand of non-standard, innovative products, and their characteristics may vary depending on individual customer needs.

The combination of the lean and the agile supply chains resulted in the development of the **hybrid supply chain** to meet the requirements of hybrid products. Hybrid products can be thought as the opposite of standard products, since those are adjustable to the customer preferences. Hybrid supply chains employ techniques such as product postponement in which the product is partially assembled, while its full assembly is pending, depending on customer demand.

From traditional to green supply chain

Supply chains have been always considered as the procedures used to transform raw materials to finished products which are then delivered to consumers. In the previous sections, it was shown that supply chains have evolved from simple, one-way processes to multi-way, complex networks involving large numbers of stakeholders. Moreover, the latter often show mutually conflicting interests and needs. The evolution of supply chains was always driven by the relevant technological advancements and market conditions. Traditionally, supply chain management focuses mainly on the maximization of consumer satisfaction and in the increase of business revenue and profitability.

However, in recent years, with the emergence of environmental issues, traditional supply chains have been held accountable for generating massive amounts of emissions and waste and resource consumption, contributing to the degradation of the environment and the deterioration of environmental status. Furthermore, supply chains are often accountable for global environmental problems, like global warming, for instance. As a result, environmental sustainability is being intensively integrated in the dominating business practices. To that end, the increasing interest of the stakeholders involved in supply chains on environmental issues led to the development of green supply chains.

The transformation process can be classified in three major groups, categorized by the source of the driving force (Figure 2.2). The key actors in these three models of evolution are the government and private corporations, who initiate the process with different motivations and aspirations as for the outcomes.

The top-down approach refers to the attempt of governments to control environmental impacts. This can be accomplished with a varying enforcement strength, from rough outlines and safe practice or conduct guidelines to strict legislation and auditing bodies, obligatory conformance to structured environmental management schemes and so on. Governments themselves are subject to control by citizens, and also by international organizations that audit environmental data globally. Accountability to these stakeholders, besides national strategic planning, is a key internal motivator in the top-down approach. This government push is described in more detail in Chapter 3. Enablers and Barriers of green supply chains, ones that analyzes the desired outcomes and also the negative side effects of opting to drive the transformation by means of a top-down approach, are further analyzed. In general, government-issued mandates are characterized by less innovation potential, stringent limits, inflexible rules and inherently high costs covered

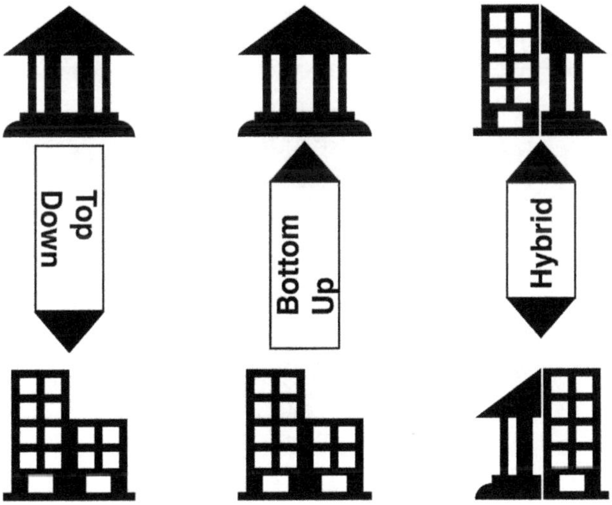

Figure 2.2 Top-down, bottom-up and hybrid transformation

ultimately by the citizens. However, it should be highlighted that the environment is often regarded as a priority for governments, even sometimes at the cost of introducing counter-productive measures.

Private corporations, under certain circumstances, are also highly interested in planning and implementing green strategy. In the case of private initiatives, the approach is labeled as bottom-up, indicating that corporations are the major drivers of change and innovation. The internal motivators of this endeavor vary from competitors' performance to corporate mission compliance. In case that all internal motivators are analyzed in depth, a common financial constituent emerges, as is expected in the case of for-profit private corporations. Analogous to governments' accountability to international bodies and citizens, corporations also are accountable to various stakeholders. One obvious group of stakeholders are shareholders, as well as suppliers and customers that are financially affiliated to corporate activities in diverse relationships. These actors impose certain demands on the organizations and influence the operations at all levels, even though they have different and conflicting priorities. On top of this situation, corporations also included society in their actor network by adopting and implementing the concept of corporate social responsibility (CSR). In the latest iteration of CSR, environmental issues emerged as a defining force compiling in a single framework[3] the three core elements and their aims to an effective business tool, as illustrated in Figure 2.3. Whether CSR is considered

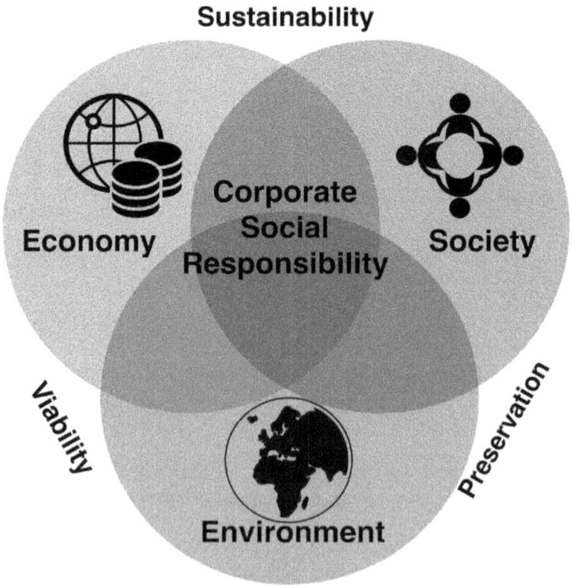

Figure 2.3 Corporate social responsibility elements

as an altruistic notion or as a strategic method to influence and shape public opinion, the environmental performance of corporates engaged in this course of action is noticeably and measurably improved. The driving factors of implementing a green strategy are in real-world conditions an array of intents and expected outcomes, covering the needs of a corporation to operate at minimum cost and maximum profitability, to altruistically reimbursing part of their earnings to the local community and, to some extent, to the global community also.

In addition to direct financial performance, indirect and preventative economic strategies are also in effect when CSR is implemented. Corporations are innately more flexible and sensitive to changes in their operating ecosystem, therefore they tend to be one step ahead of government policies. Moreover, by regularly reviewing environmentally related issues and technologies, they sustain this competitive advantage. Using these intangible assets, corporations are avoiding costly and operations-interruptive occurrences of government interference through mandates that need to be met. They also harness and employ the innovative inter-organizational potential, while at the same time are positively predisposed to externally driven innovation. Market share is a key performance index of any for-profit organization. In this light, CSR is aligned to this need by catering to the needs of "green consumers", an emerging and growing market group.[4] This market section assesses value of products not only by financial and feature/functional performance, but also by environmental performance. The added value of a green product is translated in higher status and acceptance among their peers and also as personal fulfillment, since environmentally safe practices have become an essential part of personal value systems in recent decades.

A mixture of both approaches is the most common occurrence in most cases. Such a strategy proposes a more realistic view, combining the advantages of each sector, public and private, to compile a system that is both flexible and stern. As reality suggests, legislation and practice are communicating vessels, both influencing each other. Therefore, the government can benefit from data supply of industries and their voluntary initiative to comply to guidelines and legislation and the private sector has the chance to be part of the public policy setting stakeholders. Furthermore, communication between the two sectors provides the ground to shatter pre-conceptions and facilitate productive dialog that holds the potential to formulate innovative solutions. Being able to synthesize frameworks that cater to the needs of both sectors also ultimately promotes social wellbeing.

Paradoxes in green supply chain

Although there are numerous advantages in redefining a conventional supply chain and evolving it to a green supply chain management model, there are issues that should be dealt with cautiously.[5] Corporate social responsibility can influence positively in terms of environmental performance and assist corporations to sort options by their green potential and impact. However, there are still other significant influences involved in a corporate environment. It should not be ignored that for-profit organizations have as a primary goal – often a single goal in practice – to produce profit. Therefore, their actions and decisions need to be made in accordance with this core strategic goal, often neglecting other parameters like environmental degradation.

Cost is a decisive factor for any industry. To that end, practically every business activity triggers a series of events that entail this variable. In the case of green supply chains, improvements in packaging and reducing waste flows improve the performance indexes significantly, with a noticeable effect on cost as well, albeit that some of these costs are actually part of a concealed move of externalizing costs and exacerbating pressure on the environment. Stakeholders involved in the supply chain recognize and enjoy the benefits of reduced costs, while at the same time intentionally or unintentionally ignore environmental considerations. Issues regarding environmental impact have become more popular, and the share of citizens – and consequently, officials who promote green practices – rises. However, the costs need to be assumed and governments are pressured to introduce legislation, assigning these costs to corporations. Environmental policies focusing on private personal transportation and vehicles ignore an active and widespread industry that contributes significantly to impacts to the ecosystem. It is a question of system examination boundaries that defines the positive or negative contributions of the activities entailed in the supply chain.

Supply chain strategies that have become common practice in many industries work counter to environmental safe practices. Just in time (JIT) and door to door (DTD) are both practices that introduced great flexibility and agility to the supply chain, but can exacerbate significant pressure on emissions, land use and waste production. As infrastructure needs to be developed to accommodate the increased traffic, the benefits from high environmental performance are diluted by the increase of side effects. Similar phenomena also appear due to the increase of the efficiency level of transportation. Rising efficiency reveals new opportunities in the solution space for optimized production allocation. This has enabled the spatial disaggregation of production and consumption, and brought forth longer kilometer-to-ton ratios and increased packaging requirements, negating the positive effects. In turn, the demand for transportation increases and the positive effects of technological methodical advances are negated. The case of DTD similarly increases emissions, demonstrating that besides the intensification of transportation work, the profile of usage also creates changes in an adverse way.

Reliability is also crucial for supply chain systems, since their complexity level makes them susceptible to a wide array of possible fault sources. To ensure a high reliability rate, system planning tends to favor solutions that do deliver the desired outcome, irrespective of their performance in other fields, that is, green key performance indexes. Unfortunately, in most cases, the solutions that are characterized by increased reliability – for example, air and truck transportation – are the least friendly in terms of their environmental performance.

The increase of e-commerce sales channels, in retail but also in business-to-business (B2B) uses, contributed to the significant increase of the share of road transportation. Required are also extensive warehousing facilities located adjacent to metropolitan cities, those being the main recipient of products and services procured through this channel. Although the virtualization of brick and mortar shops is considered to bring noteworthy savings of physical resources, the actual functions accompanying trading have migrated from a physical reference point, the shop, to the home or office; that is, the customer.[6] Thus, economies of scale are no longer applicable and the impact on traffic, for example, is becoming strong. Considering also that the concentration of activities is high in already congested areas, the negative effect is multiplied.

Warehousing support for all these activities and their novel modes of operation is reduced since the bulk volume of products is on the move. The energy demands and resource usage of storing products in warehouses – in contrast to transporting them from production sites or distribution hubs to the consumers – are higher and bring forth severe environmental impacts, considering the previously mentioned factors. The balance of production rates, warehoused goods and logistics needs to optimize cost-profit variables and, in the case of green models, environmental key performance indexes. Meeting all these requirements at optimized levels is virtually impossible. Therefore, endless approaches and models are practically proposed.

Chapter summary

In this chapter, we have focused on how supply chains have evolved from ancient times to the modern green supply chains we have today. Different types of supply chains were presented, and the transition from traditional/conventional supply chains to green supply chains was discussed. Moreover, this chapter explained the motivators of this transition and identified the paradoxes in green supply chains.

3 Enablers and barriers of the green supply chain

Discussion questions

- What are the internal and external barriers in green supply chain management?
- What are the internal and external drivers in green supply chain management?
- How do those influence the adoption of green strategies?

Introduction

Transitioning from traditional to green supply chain management is a process that includes obstructions as well as accelerating elements. A mix of driving and hindering forces is at play and influences the decision processes regarding engagement in green practices.[7] On one hand, the mature and widely implemented technology and mentality of conventional supply chain management have the advantage of operating already and offering their service to the production systems, contributing to the delivery of value to the consumer. On the other hand, the pressure of the imminent and current environmental issues has shifted the perspective from one based in the purely fiscal and effective to a more holistic approach integrating environmental variables and focusing on efficient use of resources, as well as associating them with corporate policies. This incorporation of environmental aspects to corporate policies is horizontally applied from economics to marketing and production to human resources.

In recent years, the world has changed significantly. Moreover, it is expected that the distribution and strength of consumer markets will be more dispersed and expanded beyond the well-established economies of the Western world. Considering that production sites, raw material extraction and processing may also be equally dispersed around the globe, the need for efficient supply chain management is clearly prescribed. The world economy needs goods to be available "at the click of a button". This imperative is partially served by the virtualization and automation of processes, goods and services. However, a substantial part of the economy relies on the physical availability of goods and services.

Beyond availability problems, the new market spaces, along with the established ones, constitute a demanding system that imposes high production volume, extended value chains and added-value goods and services. High volume is a factor that strongly affects both production and logistics, while at the same time costs need to remain constrained in order to preserve competitiveness in a global production system.

Added value is a fundamental element of products that compete in an overwhelming extended-offering ecosystem. Focus on greening the supply chain management is in full compliance both with marketing needs and corporate responsibility, as well as public policies and bio-health disciplines. Offerings that fulfill the need of sustainable production and consumption are preferred by a constantly growing sector of consumers with favorable demographic attributes. Moreover, sustainability has evolved from an ecological concern to a fundamental public policy enforcing framework and codes of conduct, as well as a measurable economic factor.

It should be noted that various stakeholders play a key role in the supply chain system. When they are internally examined, as for their capacity to facilitate the transformation to green practices and thinking, new facets of the subject are encountered.

These constituents form a prospect-filled and at the same time challenging environment that is analyzed in the following sections, equally from the obstructive and the enabling spectrum.

Internal barriers

Barriers are categorized based on their source of obstruction. Therefore, they are characterized as internal or external. A brief overview of internal barriers examined is presented in Figure 3.1, while those are analytically discussed in the paragraphs to follow.

Lack of know-how and experts

Adopting cleaner practices in supply chain management is hindered by numerous factors, with lack of knowledge and know-how being one of them. Although investing in knowledge is highly related with positive expected outcomes, such strategic decisions are scarce, and thus the potential of profitable green practices remains untapped. Lack of concrete knowledge on the materialization of clean technologies leads inevitably to the misconception that their implementation is definitely related to higher costs and risks. Even

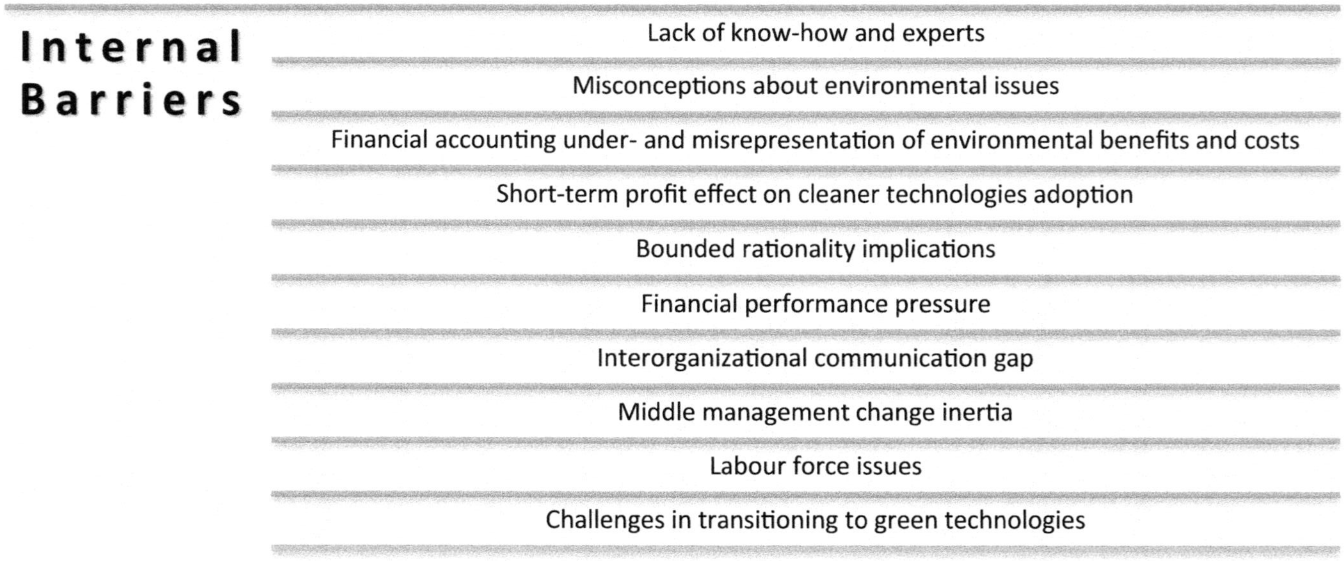

Figure 3.1 Internal barriers summary

if there is some familiarity with the topic available within an organization, in order to keep up with current scientific developments, a regular review of trends, availability and feasibility is required. In addition to outdated knowledge being counterproductive, public policy compliance may also be an issue. In this context, allocation of dedicated resources is required. This can be a significant obstacle, especially for small and medium enterprises (SMEs) that usually have limited resources. Still, even in cases where resources are at hand and allocated, the volume of information regarding the topic and the extended boundary of the supply chain field can lead to "information overload" and to the inability to process them effectively.

Misconceptions about environmental issues

Lack of knowledge is often accompanied by misconceptions about clean technologies and pollution-preventive practices. It is often seen as a "necessary evil" to comply with environmental guidelines, almost always addressing the fear of penalty and not the sense of accomplishment and reward. Viewing environmental compliance as a preventative strategy to avoid penalties leads to negating effects, for example, rendering environmental practices invisible when effective and surfacing only in a negative context when unfavorable instances occur. Regrettably, also popular is the notion that corporate environmental policy is of secondary importance. Contributing to this mindset is the fact that often the assessment of relationship between environment and corporate policy is partial and the prevention concept is vaguely comprehended. Environmentally friendly practices are also unjustly linked to higher costs, thus deeming them as an expense and, combined with other unjustified beliefs, subsequently also optional. Common in all levels of business activities, and regardless of the subject examined, is resistance to change. Businesses tend to resist changes and consider them as an annoyance and disturbance of normal and timely operation. Consequently, considering the environmental dimension of each activity in a supply chain, a system complex by nature, is a likely candidate to be viewed on an "only if necessary" basis. Public policies can also be misinterpreted as the only guideline and goal needed to follow and achieve, respectively. Poor goal setting is counterproductive in more ways than one. It can also cause fragmented focus of attention to inputs and outputs and not the complete system. This condition reflects poorly on the initiative and innovation potential of voluntary action, which will be analyzed later.

Financial accounting under- and misrepresentation of environmental benefits and costs

Supply chain activities are interrelated with financial outcomes and this is a main consideration factor at any level of decision-making processes of every organization, whether public or private, for-profit or non-profit. Detailed and accurately allocated financial data regarding supply chain activities are available to stakeholders aiding informed decisions on strategic design, operations and auditing. The "polluter pays principle" and "prevention is better than cure" are layman's terms, but they are integral parts of green supply chain management. Still, financial accounting seems to overlook environmentally related costs and liabilities, which in turn practically excludes this dimension at the decision-making level and intensifies the fragmentary environmental approach within corporate management. A financial accounting scheme should ideally factor in environmental impacts, risks, incurred costs and liabilities, even though there is some difficulty in quantifying them. Moreover, avoided costs stemming from unfavorable incidents and benefits from long-term application of cleaner supply chains are equally rarely present in balance sheets and budgets. Therefore, inaction and non-compliance seldom are accounted for. As stated earlier, lack of data leads to omitting these aspects in planning and calculating ahead, leaving potential beneficial financial effects out of the equation.

Adoption rate of environmental accounting factors are slowed down by "obscured profitability". Lack of economic evidence doesn't comply with the rational orientation of most firms that are willing to consider solutions that may offer increased profitability. Still, in an industry as highly competitive as supply chain, liquidity is always preferred over long-term returns, favoring cleaner technologies that achieve positive results in the short term.

Short-term profit effect on cleaner technologies adoption

The short-term profit mindset is a key decision-making influence for all businesses, often annulling improved environmental performance's related profits. Decision-making business executives have to account for performance and investment attractivity of businesses, and focus on indicators that serve this purpose. Environmental performance is therefore frequently overlooked in favor of dividend distribution or relative strength index that measures limited and past financial performance, ignoring future earnings or momentum. Since businesses depend on investments in order for them to be able to maintain their profitability and thus competitive position in the market, strategic financial planning takes into account factors that attract the market's interest. Environmental performance improvements frequently involve long-term investments that may worsen shallow, snapshot-like evaluations of firms.

Failure to incorporate long-term investment policies is a systematic deficiency of management tactics that has a negative impact on clean technology assessment. In turn, overall investments are based on early economic return, and this leads consequently to a lack of innovation introduction to any level of the company. Ultimately, this approach may pave the way to incremental liquidation of the firm.

Besides being healthy and potentially able to invest in cleaner technologies firms, a considerable share of the market is in a marginal economic state that reflects upon their investment strategy, as well. This part of the ecosystem may identify and acknowledge the positive potential of advances in cleaner technologies; still, their economic state prohibits them from making the change towards them. In-depth analysis of sustainability may dictate that these firms have a chance only if indeed they move aggressively and invest in clean technologies, improving their efficiency in resource utilization, transportation, warehousing activities and managerial procedures. Public policy should be sustainability-oriented to encourage this path of conducting business.

Bounded rationality implications

The gap between theory and practice is often accounted to reality factors, compromising optimal conditions. In contrast to "perfect rationality", the "bounded rationality" model is based on the assumptions that information may be incomplete, cognitive ability and time is finite. This conceptual framework, closely related to "stakeholder theory", tends to promote "satisficing" solutions, diluting managerial initiative. Stakeholder needs are a first-priority element, and therefore fine-tuning environmental performance becomes a secondary issue.

Financial performance pressure

The pressure exacerbated on businesses to perform in financial terms is ironically undermining their ability to implement cleaner technologies and unlock their profit potential. These hindering factors may have reasonable justification, or not. Investing in green supply chain management includes, to some extent, risk and uncertainties that result in skeptical consideration. Contributing factors may also be imprecise calculation, erroneously allocated environmental costs, investment strategies that lack environmental factoring-in, counterintuitive public policies – for example, waste reduction loosely monitored – or low set limits. To be able to reach informed decisions about the future requires an acute sense of predicting and assessing future liability or opportunities. Cost-benefit analysis can also be biased if it does not include dimensions that influence the system considerably. Short-term profitability, already analyzed in detail, is also a hindering factor when considering green investments, intensified by the "free-riding" competitors that operate without the "extra weight" of investment-tied capital. Costs are also hard to estimate and allocate to their true source, as they may be concealed as operational derived. Smaller firms have limited resources available, and therefore encounter major difficulties in deploying capital-intensive investments that perform based on the principle of "economies of scale". Older firms, on the other hand, have higher inertia in implementing new and radical solutions as they frequently have already invested in some sort of environmental managing solutions.

Interorganizational communication gap

"Silo Mentality" is an unwanted outcome of organizational growth. Along with firm size, departments and teams grow in size, become specialized or independent and sometimes even spatially significantly dispersed. Team mentality and belonging to a group may foster loyalty, commitment and goal alignment. However, they may also alienate the extended "family" of the company. For example, truck drivers and warehouse personnel falsely believe that their interests are contradicting. This is vertically and horizontally in effect within a firm, and ultimately has implications on all functions. Decisions about green management techniques are also affected. As an example, the introduction of a new policy may be viewed as an unwanted interference in a team's activities, thus making them less committed to the actual aim of this change.

Interorganizational rivalry may also be channeled in a counterproductive way, with groups that compete in strength and influence instead of contributing to the common goal of greening procedures or assessing new and existing technologies. Countermeasures can be introduced to change the mentality and perspective. Environmental managers should, for example, be present and more actively involved in the team formations. Communications should be omnidirectional, and therefore effort to promote both top-down and bottom-up approaches should be encouraged, although they should be monitored to avoid conflict. Inspiration may come from any level, hence the "doors" should be open and inviting for all stakeholders to present ideas. This can be boosted by encouraging the staff to embrace a green mindset and to feel part of the corporate responsibility program.

Middle management change inertia

Beyond a certain point, every activity of a business has to be split into levels. A three-tier architecture is therefore ubiquitous in the business world. Middle management is innately positioned to act as a buffer and therefore maintain balance. This comes at a cost in flexibility, and middle management change inertia and how this reflects upon the performance of the whole organization is a field of research by itself. Upper management may conclude that green supply management should be part of the organization's core strategy and take the first step in this direction. However, this is not always adequate, since it does not mean that middle management will also follow along this path. This inertia effect can be attributed to various factors. The role that middle management plays within the organization may be vital, but it is also "transparent" to some extent. Therefore, any change may be perceived as a threat, undermining its value or status. Change means also that some practices may be ceased or altered, and this is usually not welcomed by structures that have little potential to excel and their performance is almost binary on an axis of noticeable failure or "silent" success. Green supply chain management could be welcomed if it were introduced as a new practice, but also an opportunity to give middle management a chance to emerge as an efficient performing layer of the firm.

Labor force issues

On the same principle as middle management, labor force can also be an impeding factor to successfully transitioning to green supply chains. Introducing new practices and duties to the personnel usually implies assigning roles and possibly creating new positions. Should there be shortage of personnel, the implementation is handicapped from day one. Even if the personnel are readily available, the justification of allocating valuable resources, such as manpower, to environmental affairs may not be fully accepted by all within the organization.

Environmental programs are commonly introduced as an extra layer on top of firm activities, thus requiring an overhead management structure that has to be integrated seamlessly. This is unfortunately not always the case. Reasons for this may be the inability to append the program's requirements to usual activities. Even if the compatibility is not an issue, the workload of an additional management layer may not be handled by existing management potential. Changes of this nature may have a vertical effect within a firm, affecting both upper management and labor force jobs. This perceived threat may be also manifested as a resistance from the engineers to master the new skills required to transit from conventional to green techniques.

Challenges in transitioning to green technologies

Firms that are already active in the supply chain management business have already had a significant share of their capital tied-up in investments. The equipment in service has to be amortized to successfully execute past strategic and tactical planning. Extending beyond the amortization period is the useful life of equipment, frequently the essential part of its profitable use, hence the firms tend to be unwilling to retire equipment that has the potential to contribute to profitability with its value already amortized. In addition to equipment, staff may also be trained and have gained expertise in practices in service and intangible assets in the form of in-house know-how and expertise are vital to the competitiveness and sustainability of firms. Combining these factors, a larger scale change may be imminent, bringing forth radical changes in structure affecting job positions and personnel.

A complex system as supply chain, involving mature and innovative technologies, and numerous and diverse industries, products and services, is difficult to define as technology driven or mature. All wide and multidimensional systems are a combination of both. Therefore, some sectors and activities more than justify change of practices and investment of resources, whereas in other areas it is harder to relate the need to the added value that is expected to be generated. Still, considering that a competitive field such as supply management may have reached some maturity, this may also mean that the next disruptive evolution is about to happen. Firms already active have to "defend" their position against potential new players by being able to identify the right combination of mature and new technologies to achieve high and efficient performance levels.

External barriers

Consistent with the categorization scheme, external barriers are analyzed in the following section. A brief summary of them is presented in Figure 3.2.

The regulatory approach's fail points

The process of transitioning to green supply chain management through regulatory endorsement was less effective than desired. On one hand, formulating strict standards "stiffens" the system and allows little – if any – room for innovative changes to be compliant. This triggers a reaction, or to be more accurate, a lethargic response on environmental issues on behalf of the management. Since there is no profit or merit to be gained and keeping up with standards is feasible, little attention is drawn to this topic.

As already discussed, the supply chain is a complex, multifaceted system using diverse and numerous resources that all constitute the industry sector named "supply chain". As happens with practically all economic activities, involved and providing service are firms of varied sizes and by nature of work active in various industries. Regulating a system varying in every dimension is a challenge frequently faced by public policy makers, having to define boundaries and ranges of operating levels and conducts. Frameworks developed under these conditions usually favor existing firms and practices, which acts unintentionally as a barrier to new entrants or initiatives from existing stakeholders by requiring lengthy approval procedures or excessive technical proof and documentation, increasing out of proportion the cost compared with those of conventional methods.

Alternatively, regulation based on performance offers a solution less strict in terms of conceptual approach. However, issues do arise even when firms are free to choose themselves the technology mix application. Performance-based evaluation tends to bring forth the "minimum effort" mentality. As soon as minimum or maximum compliance has been reached, no further improvement efforts are undertaken. Misleadingly, regulation set ranges of

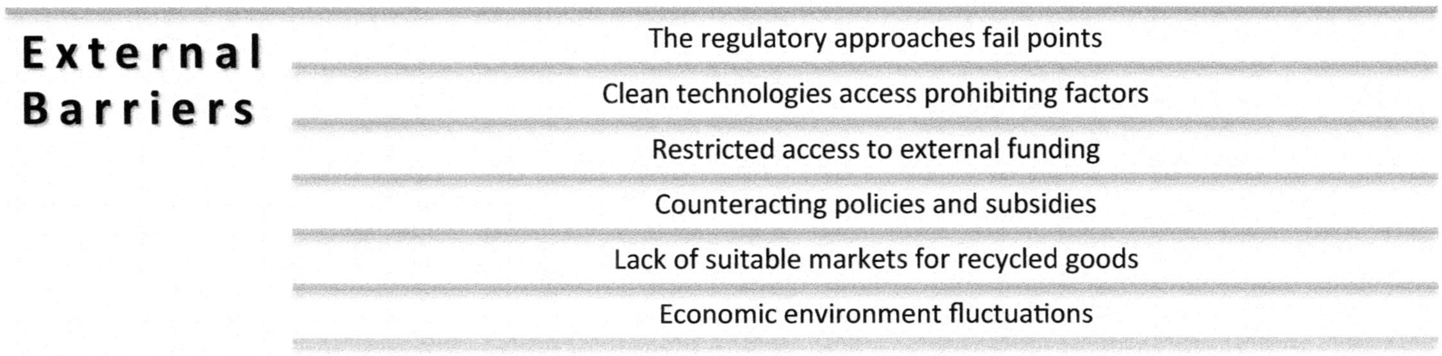

Figure 3.2 External barriers overview

operation are perceived as ideal and optimal conditions, decreasing motivation to seek novel solutions or fine-tune procedures.

In a constantly changing business environment, regulation has to keep up at the cost of stability. Every amendment means that green technologies and processes need to be examined for compliance, leading to further complexity. Besides procedural problems, every activity in business is related to cost, so therefore costs initially increase when regulation changes. In the mindset of businesses being campaigns to conquer the economic world, regulatory interventions are not welcome and perceived as hostile or a "necessary evil". Firms are usually negatively predisposed against changes. Last, focused are usually approaches to clean up polluting incidents and processes. End-of-pipe methods are usually costly, which in turn places smaller firms in an unpleasant and perilous position, giving a competitive advantage to large and resource-abundant firms. Even worse, pre-emptive solutions, preventing pollution from occurring at its source, are not promoted adequately enough, despite being a valid and egalitarian solution for all sizes of firms.

Clean technologies access prohibiting factors

The decision to invest includes risk by definition, as all actions can also have unwanted implications. Besides cost, which is an undoubtedly a significant factor, other obstacles may be encountered, making green investments less attractive or annulling initiatives altogether, effectively haltering evolution of the supply chain system. This phenomenon is intensified in the case of SMEs, which operate in adverse conditions concerning their ability to access green technologies even when they are to gain from incorporating them. Novel technology frequently is complex and highly specialized, making transitional periods harder and less productive. Even when performance specifications are high, these are achievable only in specific economic and process conditions. A vital role in greening process is substance substitution; one of the best-known examples is refrigerant working fluids. However, substitutes are not available for all potentially hazardous pollutants. Apart from absent substitutes, some of the technologies are at an early and unproven stage of maturity. On top of that, even in cases where technologies are commercially available, those may be poorly supported by suppliers or not provided as a complete and turn-key system, including both technical and managerial aspects within the offering. Competitive approaches within green supply chain management may present a dilemma between higher costs and technology sophistication levels. By rule, low-tech solutions tend to be economical, but less future-proof. Balancing cost and benefit is thus challenging at any level of firm size. Firms that decide to proceed with a substantial reform to green practices may also face quality issues, as substituting is virtually impossible to leave the system untouched in all dimensions.

The assumptions made when designing green alternatives may affect quality standards set irrespectively of environmental performance.

Technologically mature solutions, as end-of-pipe methods, are appealing due to tested and proven function, high availability and alternative supplier existence. Installation of end-of-pipe devices are low-risk moves with high reward potential. They serve their publicity purpose, green profile associations, operate successfully and are interchangeable non-interfering with the original operations as they are add-ons and not integral part of the system.

Proprietary solutions and bundled technologies can become obstacles in their retirement phases. Usually, such systems have a weighty overhead in terms of capital investment, and their architecture is not based on modularity. Therefore, incremental upgrades are not always an option to be considered, prolonging on amortization-based justification their service life at the expense of technological evolution and environmental performance. Closed systems often demand large-scale investments and transition of all interrelated subsystems to the next generation, providing little, conditional or no compatibility with older generations of equipment. On the contrary, open architecture systems using non-proprietary protocols are much more flexible and allow for smaller-scale incremental investment plans with lighter and controlled footprint on operational expenses and capital sunk.

Restricted access to external funding

Investment funding can be either external or internal, with both options being optimal choices in different circumstances. The capacity of a business to access external funding is vital for business to grow proportionally to their potential. Although SMEs represent a considerable share of global economy, they have difficulty in attracting funds for investment. Environmental investment plans are hard to finance externally, also due to the absence of specialized funding schemes that could support such ventures.

Counteracting policies and subsidies

Depending on local and global circumstances, public subsidization policy may act counterproductive to environmental performance. Energy costs are a known example of this mechanism. As energy costs are financed and partly subsidized by government, businesses can operate at lower costs, and this encourages growth and high economic performance. Unfortunately, this also demotivates and renders economic unsustainable the environmentally friendly technology development and adoption. Given the choice, corporate policy favors short-term and shareholder interests, postponing green investments to the period that they can also provide – or even guarantee – economic justification.

Lack of suitable markets for recycled goods

For firms to embrace recycling and reuse of products (packaging, equipment, components, byproducts, etc.), marketability must exist and facilitate seamless transactions and goods trading. Environmental specifications should prescribe market conditions if the recycling and reusing aspect of green supply chain management is to be viable in the long run. Despite the fact that preventative actions are preferred over managing waste or end-of-life products, the green supply chain management includes a recycling and reuse component to optimize resource efficiency.

As in every framework where change is needed to evolve towards green operations, public policy can play a defining role if the appropriate factors are weighed in. Real market needs should be considered and put into focus, and alternative uses can be introduced and promoted as long as they cater to the market needs, ensuring that market demand will exist in time, instead of attempting to impose trading actions that seem incompatible with rationality-inclined market mentality.

Economic environment fluctuations

Undisputedly, the economic environment plays a crucial role and is a decisive factor driving decisions within firms. For example, recession exerts pressure on investments, even more so in activities ambiguously regarded as environmental performance improvements. On the contrary, growth periods favor investments and corporate identity association attempts with issues such as green practices. Local- and global-scale conditions and incidents interact, forming an ever-changing economic environment.

Internal motivators and drivers

Following the scheme proposed for barriers, motivators are also divided in external and internal. In this section, internal motivators and drivers are summarized in Figure 3.3.

Environmental management systems and continuous improvement

In 1992, the British Standards Institution (BSI) first devised a structured environmental assessment standard, namely BS 7750. Inspired by BS 7750, the International Organization for Standardization (ISO) launched its 14000 family of standards. Besides these, the European Union drew up a more detailed and strict Environmental Management and Auditing Scheme (EMAS, latest version EMAS III). Although the management systems are different in many aspects, they all provide structured methods of introducing a "clean practices" mindset to improve environmental performance within companies. To achieve lasting and consistent results, a comprehensive

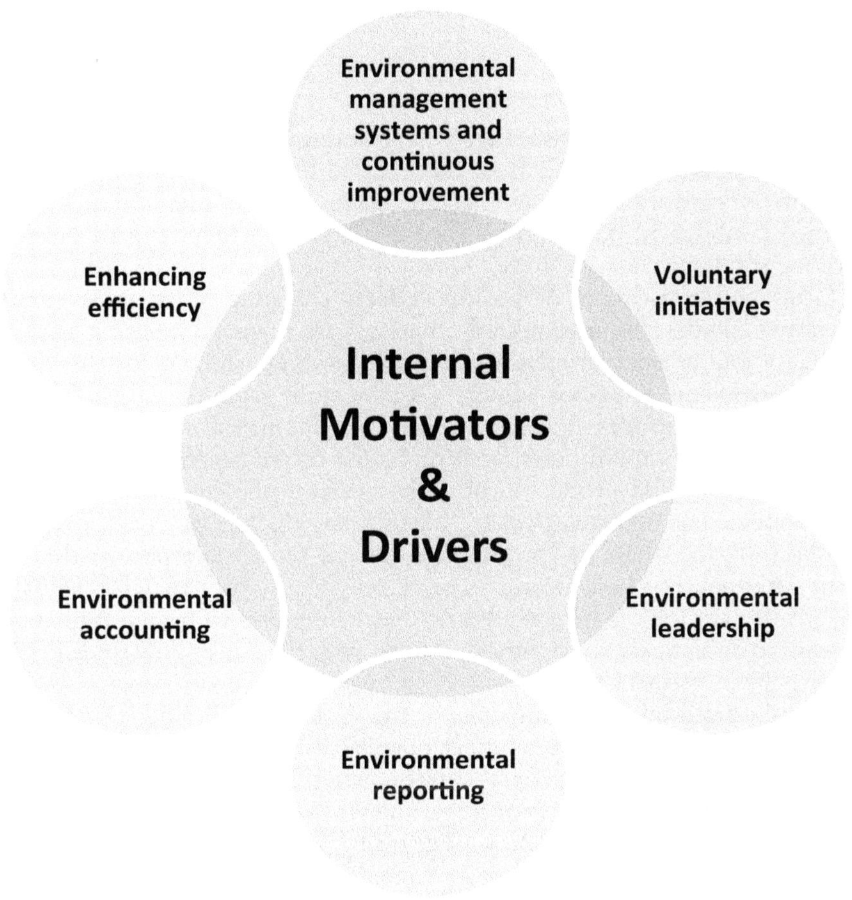

Figure 3.3 Internal motivators and drivers overview

examination of the whole organization is required in order to identify all areas that need adaptation or even total restructuring to integrate environmental management tools. An environmental management system (EMS) is fully implemented if it spreads throughout the whole organization, from core to minor activities, while it is also weighed in every planning and decision-making occurrence, positively influencing continuously the environmental performance. Firms that operate within such frameworks benefit from identifying activities through the supply chain that hold untapped improvement potential, enabling performance improvements. Also, the broadened viewing angle to the system allows firms to recognize improvement in processes beyond the narrow field of outcome ratings, or regulation.

The total quality management (TQM) framework is in many aspects similar to an EMS, since they both aim at optimizing processes in a broader rather than a financial sense only. Combinable with TQM systems, they can contribute to a structured method of organizations conducting business.

Voluntary initiatives

A notable share of green practices was introduced outside of a framework or standard, and are products of the internal or external problem-solving capacity of private initiatives. As they enjoy innate advantages – well-defined boundaries, data and test-bed availability – they tend to produce striking results.

Firms enjoy an array of positive effects engaging in environmentally friendly activities. Some expected outcomes are financial benefits, positive publicity and a sense of advancement and accomplishment for the work-force. For example, "Neon Energy", a case study from the Greek market, replaces and upgrades lighting systems to LED-based solutions. This is realized with no capital investment on behalf of the beneficiary. To finance investment cost and produce profit, savings from the energy consumption are channeled as amortization to "Neon Energy". The pay-back period is significantly less than the profitable expected service lifespan of the system. Although the beneficiaries receive only the non-financial benefits of the investment for the first few years, they can expect to step into financial savings in a relatively short period of time, with their cash flow undisturbed during the investment period.

State-inspired initiatives also play a significant role. Besides setting policies, governments can also support positive driven actions, as awards or publicized listings of exceptional environmental performance, etc. Simultaneously to financial savings for governments, absence of costly inspection structures etc., the state's role is also transformed from a penalizing and policy enforcing player to that of a rewarding and assisting one.

Other drivers for voluntary action may include peer pressure for firms that are not pioneering in environmental performance, in contrast to their competitors. This may also be the result of chamber and associations memberships that also push their own environmental policies to comply with a green profile for the group. Also, when competitors do engage in greening their supply chains and exploit the positive outcomes, firms need to respond in order to keep up and defend their market share. Last, there is also a "trickle-down" effect to be noticed, as was the case with the introduction of ISO 9000 quality systems. As high-profile organizations integrate the framework, this also is manifested in their suppliers or collaborating requirements, steering whole industry branches towards the same direction.

Environmental leadership

The term "environmental leadership" refers to the strategic decision to excel in environmental performance. This is a major view change from environmental issues as part of the firm's activities to environmental performance becoming a key performance indicator on the same level as market share. Assignment of high priority also translates to involvement at all layers of a company, which in turn inspires a universal sense of achievement and progress.

Personal investment is an important factor in committing to programs that demand radical changes in practices and everyday actions and unifying, team-building actions, as leadership goals are encouraging the workforce to participate. This is intensified in the case of environmental goals, as corporate activities reflect usually also in the local area that the workforce leads their private life, too. As members of the firm that strives to improve the local conditions of living, the workforce's self-esteem also rises and a feeling of accomplishment and pride at a private/personal life level is also induced. Local authorities also view positively companies that display responsibility, accountability and respect towards their hosting surroundings. Gains are brought at financial level, too, as these actions are publicized and financial stakeholders are influenced by them.

Environmental reporting

Companies with active environmental management plans communicate their performance results to various stakeholders. Dissemination actions are part of the extended value to be exploited by engaging in green programs, but the collection and interpretation of data itself is a potent internal auditing procedure. Managers responsible to produce this report need to examine the firms' activities within a period and have a panoramic overview of actions, results and points of interest. As the system is viewed holistically, bottlenecks can be identified, together with unexploited areas of potential gains or underperforming measures. Less noticeable at local level, performance differences are noticeable at a system-level view.

Currently, practice methods include – among others – eco-balance sheets, recording inputs and outputs for the sum of business activities, performance indicators, carbon footprinting for services provided and goods produced. A common and quantified scale of performance aids in setting measurable goals, monitoring and assessing efficiency and progress. Another advantage of quantified data is that they provide a common comparison base on which environmental reporting extended in other corporate activities may be evaluated.

Environmental accounting

Corporate or even smaller schemes have a detailed and extensive financial accounting system in service. Tracking environmental performance and impacts also generates financial data that needs to be merged in the existing financial accounting system to be included as a decision factor when financial data are needed. Integrating data is a challenge, but also an opportunity to fortify the value of green practices and allow businesses to harvest the financial benefits related to them, increasing their profits and supporting their sustainable development.

Accounting data are generated from all inputs and outputs of the organization, and in terms of greening processes energy and materials are a primary source of data. By nature, managerial-related activities are translated into monetary terms and such information regarding environmental management is included in the accounting scheme. Life cycle assessment is a widespread method used to evaluate impacts of an offering's existence through the whole period, starting from conceptional and design work to end-of-life handling. This approach enables organizations to assign the environmental cost to periods, and therefore rank them accordingly in terms of contribution. Life cycle cost assessment is processing impacts on a financial basis and quantifying them into monetized units, easily merged with financial accounting systems. Furthermore, the existence of an organization and the activity within its environment brings forth environmentally related consequences. Those are accounted in environmental impact and externalities cost assessment.

Informed decisions have higher rates of success and financial dimensions of supply chains are ranked among the highest priorities; hence, environmental accounting is a tool potentially providing competitive advantage to organizations that harness its power.

Enhancing efficiency

To some extent, effective supply chain management is identical to green supply chain management – and therefore, enhancing the organization's efficiency through transitioning to green management is a valid and promising approach. Focal points for improvement change – albeit, the desired outcome remains the same in its core – to ensure that the firm will be profitable and sustainably developing. Implementing a holistic framework, technologies and management synced to optimize green performance and efficiency as defined in each individual supply chain is improved. Although such transformations require time and contribute in small increments to a company's turnover, costs related to environmental dimensions are usually high and therefore a promising area of improvement. Usually related with long pay-back periods, in some cases green management introduces also rapid or even instant pay-back periods. Numerous environmental performance enhancements involve change of everyday habits (e.g. regulating air conditioning at a reasonable temperature or using timers to ensure that they are turned off), translated to zero-cost activities despite the fact that profits may occur instantly. A scalable plan could be devised that under conditions is self-financed.

External motivators and drivers

In this last section, the external motivators are enumerated and briefly described. The overview of motivators and drivers identified are depicted in Figure 3.4.

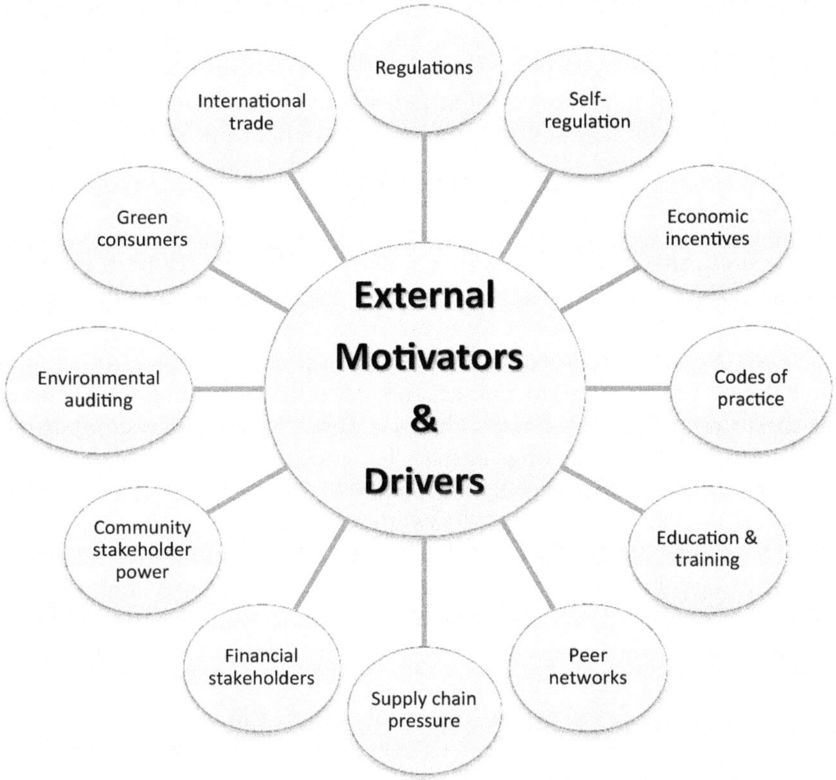

Figure 3.4 External motivators and drivers overview

Regulation fostering pollution prevention innovation

Traditionally, governments regulated activities through legislation dictating actions, defining permissions, assembling control bodies and setting penalty systems for incompliant conduct. In recent years, however, some innovative attempts have emerged. An influencing factor was former US President Bill Clinton's environmental agenda, that brought the "Reinventing Environmental Regulation".[8] Focus was set to private organizations that develop custom solutions to deal with environmental management. These attempts tend to be carefully planned to be profitable, innovative and feasible. Another major shift in framework design is that the viewing angle has changed to system-level instead of specific events. Hence, pollutants are considered as part of the whole system's input-output procedure. This change has a two-fold effect on implementation. First, the whole supply chain is examined in the permitting procedure. More specifically, all activities related to the organization are included and examined under a preventative microscope. Second, pollutants are categorized, and their limits are set, according to existing legislation/regulation. This results to a system that allows central and strategic planning, functioning pre-emptively, reducing costs and promoting inter-organizational technology development initiatives.

Self-regulation

Regulation drafting is a costly process that needs to satisfy both the public as well as the private interests. Achieving this balance is almost impossible when both parties are not involved from the earliest stages of development. Self-regulated sectors, when designed on a sound and sincere basis, can balance financial performance and compliance with greater success. The development of one-sided policies is subject to reduced compatibility with common practices, often narrow in scope and lacks consideration of real environmental improvement capacity of organizations, whereas self-regulation involves a two-sided dialog that enables the conflicting interests to reach an acceptable compromise. It also opens a communication channel that brings the two parties closer and allows for deeper understanding of other-side requirements and mentality, making future reviews of regulation easier and more efficient. It also greatly increases the adoption and compliance rate, as limits and requirements are set at levels that serve both public policy makers and private for-profit organizations. As negotiated self-regulation is drafted under consensus in every step of development, time and cost expenditures are reduced and adapting periods are minimized.

Economic incentives

To promote green supply chain management, various financial incentives may be offered. Frequently used incentives include tax deductions, public subsidization for investments, etc. On the other hand, subpar performance may also be penalized through environmental load compensation charges or increased tax coefficients. Since financial performance is closely monitored, credit or debit will be recorded and reported to managers of all levels and in turn environmental policy-induced monetary flows will be a parameter of consideration. In terms of awareness, economic incentives perform adequately; however, in terms of steering development towards sustainable solutions, their effectiveness is not clear. To overcome makeshift attitude and short-spanned solutions, governments can subsidize capital-intensive anti-polluting investments and environmental management systems integration, making large-scale plans more attractive to businesses.

Besides encouraging investments in technological solutions already available in the market, governments may opt to invest into promoting research and development of novel methods. Summarizing possible economic incentives in a non-restrictive list:

- Taxes and other charges on environmentally damaging activities and materials.
- Energy consumption rate taxes and carbon emission-equivalents taxes.
- Pollution permits.
- Tax deductions for green investments.
- Subsidized equipment purchases and consulting in green technology and management.
- Subsidized research and development (R&D) voluntary initiatives.
- Funding basic and applied research in mixed consortiums (academic and private).

Codes of practice

Issuing codes of practice instead of regulation may increase participation and adoption significantly. This can be attributed to the following advantages:

- Specialized for the sector, increasing interest to making the code part of the toolbox.
- Adaptable and profitable methods.
- Market-related solutions.
- Compiled in terms already in use, increasing digestion and incorporation rates.

Even though codes of practice cover all the prerequisites for success, a complementing stricter regulation can be used to cover for non-conforming players and discourage intersectoral free-riding attempts.

Education and training

To sustainably introduce changes, it is always a good practice to infuse knowledge and training into the supply chain system. Educated and trained personnel are expected to impart their practices and knowledge to their collaborators. In this light, improvement in environmental performance will become a fundamental component of a firm's mindset. To lead an effective campaign, the desired results have to be achieved by addressing the correct audience. Targeting the correct audience is just as important as delivering the correct message, and the process is optimized when the audience members are positively predisposed and already interested in expanding their knowledge in green technologies and management.

Plotting the path to a sustainable future entails preparing generations of managers, workers and consumers who are educated in every step to act environmentally responsible. As one can easily understand, this is a long-term goal extending even well beyond business terms of long-term planning.

Critical to the success of every educational attempt is the dissemination of results – and under this light, information must be open and accessible. Data and techniques can be incorporated into existing databases or new sites and organizations may be founded and launched to deliver knowledge to interested parties. Larger-scale actions as demonstration sites and model services may require higher expenditure but in return offer hands on knowledge and instant proof of concept to observers. Their use may also be extended to be part of an educating and training program as full-scaled test beds.

Peer networks

Conventional dissemination of knowledge and results involves a centralized system of knowledge repositories that contain the total of information to be. In a decentralized model, information is being communicated on a peer-to-peer(s) basis. This model requires that stakeholders of the supply chain industry form organizations that will foster their networking and communication efforts. Usually groups form on activity criteria, compiling uniform teams of e.g. suppliers of office ware, but there are benefits in organizing according to spatial criteria since environmental issues can be strongly spatially dependent.

These networks can also function as hubs of innovation, central consulting commissioning, a reference point for government-derived information and, in the case of multidisciplinary networks, an exchange platform for novel synergies.

Supply chain pressure

In a supply chain, not all stakeholders are equal in size and/or importance. In most cases, larger firms have significant power over smaller ones and may use it to drive development to a certain direction. A manifestation of this relationship are supplier requirements that include active environmental management systems.[9] This may extend to the point that external assessments are carried out prior to committing in long-term partnerships or as part of routine supplier evaluation for their own EMS. As stated earlier, usually SMEs are the recipients of this pressure. The latter are difficult to regulate and control due to their number and diversity, making regulatory work for governments costly. However, in the case of supply chain service providers and their clients, the auditing and self-regulatory mechanisms are less needed as environmental performance is tied to core economic activity. Still, special care should be taken to ensure that imposing an EMS to smaller firms is not perceived as hostile and workload-increasing action, but on the contrary is positively associated with all the benefits that SMEs can expect from increasing their environmental awareness. Actually, smaller firms can benefit greatly and are instinctively familiar with concepts such as efficient resource utilization or zero-waste policies. Governments are also one of the biggest recipients of goods and services and in this role, they can also promote an environmental agenda.

Financial stakeholders

The business ecosystem includes financial stakeholders who, similarly to the supplier-buyer relation, exercise noteworthy influence on their clients. Enterprises rely on banks for financing, insurers for financial safety and investors for raising capital. Should these stakeholders require compliance to environmental performance standards, enterprises would comply and implement an EMS in order to fulfill requirements and remain financially active.

Besides setting prerequisites, environmental performance can be interpreted as a sign that the firm is following contemporary trends and is informed to latest standards. This implies also that a firm is "healthy" enough to review its activities beyond the narrow profitability spectrum, but also setting the grounds for possible upscaling. Besides being ambitious, environmentally friendly businesses are also less likely to suffer from penalties given to underperforming businesses.

Community stakeholder power

Corporations have undergone various transformations in structure and awareness stages to reach a mature state and include corporate responsibility as an integral part of their areas of interest. Equivalent to carbon footprint, community footprint is a pivotal point at all levels of planning for management. Concerns about environmental impacts of the supply chain activities need to be considered and be alleviated to strengthen trust to corporate good-willed intentions. Actions related to corporate responsibility should aim to act pre-emptively and build the desired corporate image.

As previously mentioned, awards and peer networks can aid in publicizing positive actions and environmental performance advances of supply chains active locally. Awards can also initiate public conversations, stimulating community interest and educating them about environmental issues beyond the confined scope of corporate activities.

Environmental auditing

Externally commissioned auditing services on environmental performance should be looked at in a positive light. Reports of environmental audits contain information that can enhance financial performance. Provided that a firm already has an environmental managing system in service, auditors can shed light on processes and reveal weak spots, opportunities and potential threats. In more detail, auditing can aid organizations to stay compliant with current legislation and avoid fines and liability. Also, operational planning can benefit from a holistic review to identify optimal relations, possible improvements and in general benchmark current performance levels. Material and energy flows are recorded, and their optimizing potential can be evaluated. The purpose of transitioning to green practices serves two goals, as elaborated earlier, both environmental responsibility and profitability. Environmental auditing is a valuable tool contributing to the successful pursuit of both goals.

Green consumers

Public awareness of environmental issues has been rising since the 20th century, and this led to the assembly of formal and well-known groups of private individuals who share a common concern about their consuming behavior and consequently about the industries involved.[10] These concerns are often voiced publicly, exerting significant pressure on providers to comply since these groups have a wide membership base and also are favorably viewed even by non-members.

Firms that do operate within environmentally responsible guidelines rank higher in consumer preferences, project an image of responsible corporate citizenship and commitment to community. A positive public image and marketing benefits are invaluable assets to any business and a major aid to sustainability.

In the case of green supply chains, this is not a direct marketable asset, as these operations are usually considered as behind-the-scene processes. Still, when communicated appropriately, such initiatives can be actions of high added value to the offering of any retail product or service.

Since green orientation is part of formal education at all levels and environmental issues become household terms and affect everyday practices of individuals, every consumer may be considered green to some extent. This might subconsciously even apply to procurement managers, and thus have power over B2B relationships also.

International trade

Goods, even services, are in constant movement within a globalized world economy. Moreover, geographically dispersed stakeholders influence remote partners, forming universal trends and practices. Whether in a business-to-business or retail framework, relationships require standards of conduct and offer to ensure consistency, comparability and quality measurements. Additionally, international trading has to also comply with recipient legislation, increasing complexity of transactions. Global market transaction volume exceeds local numbers by far, and therefore holds the potential to enforce green practices more efficiently. Supply chain management plays a vital role at this scale of trading, and businesses globally active are frequently early adopters of such standards. It is in their best interest to have consistent protocols of action throughout their branches, even when mandative legislation has not reached some of them. This leads inevitably to "trickle-down" effects and local businesses also follow their lead. In general, global level pressure is a tool used to push green issues even higher in the agenda of trading organizations.

Chapter summary

This chapter identified and critically analyzed barriers and motivators in green supply chains, and characterized them as either internal or external factors.

Internal barriers include lack of know-how and experts, misconceptions about environmental issues, financial accounting under- and misrepresentation of environmental benefits and costs, short-term profit effect on cleaner technologies adoption, bounded rationality implications, financial performance pressure, interorganizational communication gaps, middle management change inertia, labor force issues and challenges in transitioning to green technologies.

External barriers include regulatory approaches fail points, clean technologies access prohibiting factors, restricted access to external funding, counteracting policies and subsidies, lack of suitable markets for recycled goods and economic environment fluctuations.

Internal motivators include environmental management systems and continuous improvement, voluntary initiatives, environmental leadership, environmental reporting, environmental accounting and enhancing efficiency.

External motivators include regulation fostering pollution prevention innovation, self-regulation, economic incentives, codes of practice, education and training, peer networks, supply chain pressure, financial stakeholders, community stakeholder power, environmental auditing, green consumers and international trade.

4 Inbound logistics (green procurement)

Discussion questions

- Why is procurement critical for the optimization of supply chain efficiency?
- What is "green" procurement, and how is it differentiated from traditional procurement?
- Which are the factors that contribute to increasing interest in green procurement?
- Which are the tasks of green procurement that constitute the green procurement life cycle?
- What are the benefits and the barriers to a broader adoption of green procurement?

Procurement is a critical business function; it is indicatively cited that in the private sector a material purchased has a value equal to 50–70% of its sales cost. Furthermore, public procurement accounts for approximately 20% in most of the countries. Improvements in the procurement process will, on the one hand, bring about a decrease in cost, thus having a positive effect on cost, while simultaneously increasing the offered level of service by offering high quality products and services and, consequently, satisfied customers.[11] Procurements have significant effects on the other operating functions of an enterprise's logistics system.[12]

Of course, all these functions, such as the production, processing, storage, transportation, distribution and use of the products affect the environment, since they consume energy and raw materials, and emit gases contributing to climate change and/or other harmful pollutants. Green procurements related to all these products and services – the production, processing, storage, transportation, distribution and use thereof – have less of an environmental impact compared to the corresponding conventional products or services.

Terms like green (sustainable, social) procurement (purchasing, sourcing) are used interchangeably, and in order to appreciate their benefits, the following questions arise:

- What are the differences between traditional and green procurement?
- Which are the factors that contribute to increasing interest in green procurement?
- Which are the tasks of green procurement that constitute the green procurement life cycle?
- What are the benefits and the barriers to a broader adoption of green procurement?

The following sections will answer these questions.

Defining green procurement

Many different definitions have been offered for green procurement, such as environmentally friendly, sustainable and socially responsible procurements. The existence of so many and different definitions stirs a confusion with respect to the precise definition of green procurements and what it is that such procurements include. What essentially creates this confusion is the existence of many aspects for the assessment of green procurements, such as environmental, social, economic and others, as well as the fact that initiatives emphasize and target different objectives, such as recycling, energy efficiency, decrease in the use of water, decrease of waste and harmful chemicals for the environment and health, etc. This fact also causes problems when defining policies and developing the corresponding practices and methodologies.

However, all definitions converge on the following: the implementation of green practices in the process of procurements. The procurement, for example, of products/services with specific characteristics/specifications relating to their production, storage and distribution with low energy, water, etc., consumption, or products that are manufactured out of recyclable, non-toxic materials, etc., and, moreover, the measurement, recording and monitoring of all factors negatively affecting the entire life cycle of the product or service. This process is a laborious and difficult one due to the interaction of such factors and the difficulty one has accessing accurate and reliable information, while it also accentuates the importance of green labels.

The following might serve as a general definition:

> *Green Procurement is the purchase/acquisition of environmentally friendly – that is, green – products or services.*[1]

This definition leads to the following conclusions:

- Green procurements seek green/environmentally friendly products such as raw materials or semi-finished products with no environmental or social impact. When referring to green products or services, we mean those where one or more green practices have been implemented in one or more points of their life cycle in order to minimize CO_2 emissions, decrease water consumption and the polluting of water, dampen the effect of hazardous chemicals and decrease the excessive consumption of raw materials.
- Green procurements focus on the social and economic aspects of procurement, such as sustainable development. Namely, it must include in the selection criteria of appropriate suppliers, such as price, reliability, the supplier's reputation, and social and environmental criteria, issues and effects.
- Green procurements aim to limit the burden to the environment directly related to the choices and activities regarding the procurement cycle for a business.
- The following relation results from the previous: In many cases the formula *lean = green* stands. This means that green procurement implies not only purchasing green goods, but also products that are really needed.
- The objective also includes that executives, employees but also the business partners of a business adopt a mentality with respect to procurements and product purchases which will characterize their behavior in their daily lives.
- With respect to services, the objective is for them to have green credentials, which usually translates to attempts to decrease the environmental impact at the time services are rendered.

Green procurements involve and interest both roles of the procurement process, namely suppliers and customers. They do not simply relate to the transaction between the seller and the supplier, however, but rather focus on the acquisition (purchase) of products and services. Namely, they do not include the placing of orders with the selected suppliers, the payment thereof or the transportation of the products and their collection by the customers.

They principally relate to procurement, which is a broader concept including more works preceding as well as following purchasing, such as market research, negotiations, selection of the suitable supplier, pricing and closing deals, as well as monitoring the supplier's performance and the management of emerging risks. All these works comprise the procurement cycle, which will be presented in the coming sections.

Moreover, they also related to strategic sourcing, a broader, more transformational process, performed at a strategic level. Strategic sourcing takes the procurement process further, examining the whole supply network, its linkages, and how they impact procurement and purchasing decisions. It can be defined as a systematic, long-term and holistic approach to obtaining current and future needs of an organization. Therefore, the focus in sourcing includes value creation, risk and uncertainty in the supply chain, as well as the overall responsiveness and resilience of the supply chain, and furthermore, the establishment and overall management and optimization of supplier relationships and alliances – even the joint planning of products and acquisition of the strategic sources.

Of interest is also the definition offered by the organization Eco-Buy, which attempts to include the three principal outcomes of any green-sustainable practice:

> Green purchasing and environmentally preferable purchasing relate to the consideration of environmental impacts and costs in the procurement of goods and services, whereas sustainable procurement considers the social, environmental and economic implications of procurement.[13]

The outcomes of green procurement can be considered as the factors that contribute to its increasing interest. For more than two decades, many big multinational companies – but also and small and medium enterprises (SMEs) – have run many initiatives.

Factors that contribute to increasing interest in green procurement

Green procurement has different goals than does tradition procurement. Traditional procurement focuses on the "value for money" aspects of procurement. On the contrary, the aim of green procurement is to integrate environmental considerations into the procurement procedure thus aiming to reduce its adverse impact on human health, social conditions and the environment, thereby saving money for organizations and the community at large. These benefits have compelled companies in both public and private sectors to adopt green practices and use green procurement practices and techniques.[14,15]

On the internet, the reader can find numerous best practices for the application of green procurement techniques (see Table 4.1).

But many public bodies have also implemented schedules for the procurement of products and services with a reduced environmental impact throughout their life cycle (Table 4.2):

The results of the initiatives listed can also be the benefits (Figure 4.1) ensuing from the implementation of green practices in the procurement process of companies/organizations:[2]

- Reducing harmful emissions, waste generation and the use of natural resources; improving air and water quality and preventing pollution by supplying/procuring low carbon, energy/water-saving non-toxic products and products with recycled content.

Table 4.1 Best practices for the application of green procurement techniques in enterprises

Company	*Green procurement practice*
Apple Inc.	Auditing and continuous monitoring of the practice of green policies by suppliers.
BT Group	Development of close collaboration with suppliers/ manufacturers of the products the company distributes to the market, so as to ensure that the former design and manufacture products while decreasing greenhouse gas emissions.
Hewlett-Packard	Requires its key suppliers to comply with national and international regulations by asking them to use self-assessment questionnaires, and also to participate in audits.
Jaguar Land Rover	Requires from its key suppliers to disclose CO_2 emissions and closely collaborates with them for their uninterrupted monitoring and control.
Nokia-Siemens Networks	Implements a specialized software application to measure the carbon footprint of its main supplier while simultaneously training them and working with them to design green and corporate social responsibility practices.
PepsiCo	Offers training seminars to all suppliers on issues relating to the advantages of green practices as well as the optimal management of emissions and rational use of chemicals.

Table 4.2 Best practices of the application of green procurement techniques in the public sector

Country/Organization	*Green procurement practice*
China/Lists for Environmental Labeling Products	Development of lists with products which have been granted Environmental Labeling certified by competent certification bodies.
France/Union of Public Purchasing Groups	Establishment of a methodology and the corresponding tools whereby various public-sector services can measure how green their purchases are.
India/Indian Railways	Demanding key suppliers to design, develop/replace energy inefficient incandescent lamps for their lighting needs with energy-efficient but, at the same time, of higher initial cost compact fluorescent lamps.
Italy/Consip	Analysis of local markets in the country using historic data, as well as questionnaires filled out by the companies' executives, to self-evaluate and measure the degree of implementation of green practices.
Sweden/Sustainable Public Procurement	Development and use of the sustainable public procurement (SPP) tool, an internet-based application whereby procurers can choose among sustainability criteria for various goods, services and work contracts.

Benefits from the implementation of Green Practices	Reduction of harmful emissions, waste generation and the use of natural resource.
	Incremental savings of cost, by rationalizing superfluous purchasing.
	Value creation obeying the commands of society and the market relating to the procurement of environmentally friendly products.
	Improvement of work conditions.
	Creation and offer of green products to the market.
	Establishment of a framework of engagement, transparency and ethics between the company and its suppliers.
	Support of local producers (local industry) who apply green production practices.
	Support of sustainable strategies.

Figure 4.1 Benefits from green practices in procurement

- Incremental savings of cost by rationalizing superfluous purchasing. Green procurements can be considered as the tool for cutting costs related to the acquisition of products when required, at the lowest total cost of ownership (TCO) (the purchase price of an asset plus the costs of operation) and savings through life cycle costing, as well as lowest risk to the supply network.
- Value creation obeying the commands of society and the market relating to the procurement of environmentally friendly products (e.g. confronting with global environmental problems such as global warming, acting as an example of engagement). This leads to acquiring recognition and reputation which can be supported by certificates and awards, as well as, complying with multilateral environmental agreements and labor conventions.
- Improvement of work conditions in issues relating to security, safety, labor standards and hygiene.
- Creation and offering to the market of green products, products produced out materials which are, in turn, procured from suppliers who have adopted environmental practices and methodologies in the design and manufacturing of green products.
- Establishment of a framework of engagement, transparency and ethics between the company and its suppliers, while the performance and transparency of procurement procedures and the organization thereof is improved inside the company itself.
- Support of local producers (local industry) who apply green production practices, and which contributes to generate "green" jobs.
- And, of course, support of sustainable strategies.

All these attest to the great importance of green procurement for the modern enterprise, but also to the need for their integration with the other functions. The objectives, besides reducing environmental impact, remain the same: reducing ordering times, reliable and timely delivery, completeness of orders, faster response to customer demands and ability to respond to special requirements.

Green procurement life cycle

Strategic sourcing is a systematic and integrated process. It includes specific and distinguishable stages which are performed, while at the end of each stage a specific output ensues, which also serves as the input for the next stage (Figure 4.2). This process is repeated every time (that is, cyclically) when the enterprise wishes to procure environmentally friendly products or services. It is for this reason that the process is called "Supply Cycle".

Preparatory stage: choosing the green sourcing strategy. The adoption of green practices can be incorporated in an integrated procurements framework which serves the strategic objectives of the enterprise and one whereby

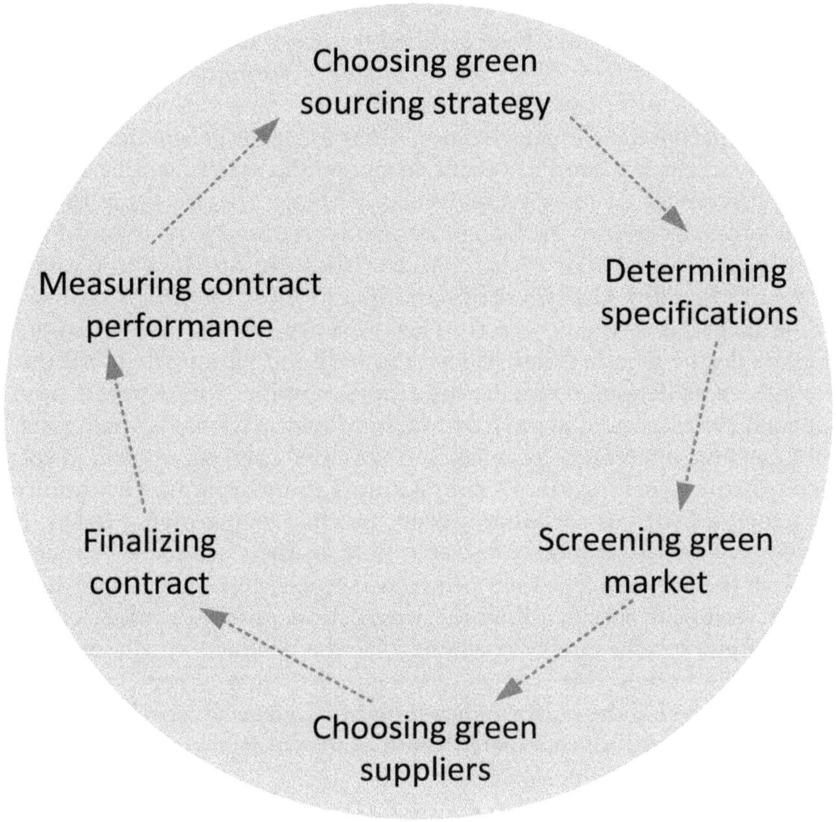

Figure 4.2 Green procurement life cycle

all departments/operating areas of the enterprise participate. The principal competences of a procurements management department mainly relate to the management of all stages in the procurement process (i.e. of the procurement cycle) in collaboration with the other departments (financial, production, supply chain and logistics, quality, marketing, etc.). Moreover, the collaboration also extends to decision-making, which principally relates to the designation of the real needs of the enterprise for green products and then regarding the green products (materials, equipment) or services that will be acquired, the determination of their specifications, the budget, the decision to make or buy, to use local or international supply chains, to have a small or extended supply base, etc.

Moreover, when selecting the stages, the entity will also take account and implement environmental criteria or considerations. If, for example, the process is an open one, then such criteria will be applied only with respect to the final selection of the suitable suppliers. In case of a closed process, they will be applied earlier, since the call for bids is send only to a limited base of suppliers who have been certified for green initiatives.

Second stage*: *determining the technical/environmental specifications. The main activity of this stage is to establish the specifications for the products to be purchased. The organization wishing for the procurement of green products can request products made from specific materials, that contain a certain percentage of recycled or reused content, that are made following specific procedures or to exclude or demand a reduced percentage of harmful for the environment or health characteristics, etc. Specifications establish the required quality characteristics and qualities for the products that will also be included in the contracts. They also assist enterprises – candidate suppliers decide whether they will submit bids and participate in the procedure as well as helping designate the criteria employed in the evaluation of suppliers. References to standards (international, national or industrial, as well as technical reference systems) and eco-labels play an important role in green procurements for many manufacturers to incorporate environmental characteristics such as carbon footprint, product or packaging re-use, consumption of energy and/or water, etc., both in their productive process as well as in the packaging of their products. Similarly, businesses may include in their technical specifications the production processes and techniques (for example, for the energy consumed for the production of a product to originate from renewable sources of energy or the use of recyclable materials). There are also the performance-based or functional specifications that describe the desired or anticipated result, and which also recommend indices for measuring their performance.

> The internet site of the European Commission (http://ec.europa.eu/environment/gpp/eu_gpp_criteria_en.htm) is most useful and interesting, as there one can find the recommendations for the technical/environmental specifications for many public-sector product/services categories, such as computer and monitors, copying and graphic paper, food and catering services, gardening products and services, furniture, cleaning products and services, etc. According to the site: "criteria aim to reach a good balance between environmental performance, cost considerations, market availability and ease of verification".

Moreover, this second stage includes an assessment of the cost for acquiring the materials, while the time-schedule relating to their demand and acquisition is also planned.

Third stage: screening of the green market. The third stage includes screening the market to find the suitable suppliers. The key sources of information are:

- Records/database for suppliers kept by the company and which includes historic data relating to suppliers (from previous collaborations or expressions of interest).
- Internet search, specifically one carried out in specialized internet sites where one can find lists of certified suppliers.
- Sector exhibitions or meetings of international and national organizations, institutes, etc.
- Bibliographic research into specialized trade periodicals, annual reports by consulting agencies, etc.

A critical factor for the success of the screening process is the existence of updated but also open access databases for the available in the market environmental choices of products or services, their environmental effects, their lifespan, as well as their price levels (an example is the site http://ec.europa.eu/ecat/ which includes the EU Ecolabel Product Catalogue where information on environmentally friendly high-quality products is kept).

> The EU Energy Star program is managed by the European Commission and follows an Agreement between the European Community (EU) and the Government of the United States (www.eu-energystar.org/). The principal deliverable of the program is the development and updating of the EU ENERGY STAR database, which allows the executives of organizations and enterprises to pick the most energy efficient and qualified office equipment under the newest active specifications.

After the completion of the market research for green suppliers, the organization announces the request for proposals as follows:

- Request for information (RFI) is an open enquiry for a company, which is seeking information and understanding of its needs; it is used when "you think you know what you want but need more information from the market/suppliers". An RFI is usually be followed by a request for proposal or a request for quote.
- A request for proposal (RFP) is a request for specific solutions to the company's sourcing problem (including terms, conditions, product specifications, delivery, packaging and labeling options of the required products) and it is used "when you know you have a problem but don't know how you want to solve it".

- A request for quote (RFQ) is used when "you know what you want but need information on how suppliers would meet your needs and how much it will cost"; it gives the opportunity to potential suppliers to competitively cost the final chosen solutions.

An RFP or RFQ incorporates policies, documents and guidelines, as well as green standard clauses. Potentials suppliers respond to them by providing their sustainability and policy mission statement, past initiatives, awards, disclosure of environmental violations, etc.

> On many occasions, tenders are used in case the purchaser knows precisely what good they want and is looking for the best price to deliver it demanding of potential suppliers to disclose their profile, the technical specifications for their products as well as, of course, their price.

Fourth stage: choosing green suppliers. In this stage, the interested suppliers send their bids/tenders/offers. The procurements department or the committees appointed by management to assess offers and select the appropriate suppliers based on specific evaluation criteria. Such criteria may be inclusions and exclusions criteria. Moreover, this stage includes the determination of their weight/importance. These criteria differ from the technical/ environmental specifications cited in the second stage. Their principal difference is that specifications are assessed on a pass/fail basis, while criteria are weighted and scored accordingly. Of course, both are assessed by the client's executives. Furthermore, they differ from those employed for simple procurements (such as, for example, after-sales services, the quality of the products and services, capacity and availability of infrastructure, equipment and systems, reliability with respect to service times, adjustment to the particular demands of clients, etc.).

Researching the relevant bibliography (sources including web sites, eco-label schemes and tender documents) selection (or award) criteria are the following:

- Cost of green products (1).
- Use of green practices for the production of the products, such as use of renewable energy sources, use of recyclable materials, low energy and water consumption, low levels of waste, CO_2 and greenhouse emissions, etc. (2).
- Compliance with applicable international and national environmental laws.
- Compliance with international, national and business environmental standards (such as the Environmental Management and Audit Scheme – EMAS and the ISO 14001) (3).

- Competences and references related to environmental aspects.
- Experience in environmental products.
- Infrastructure and technical resources for environmental protection.
- Use of special software applications that ensure the conformity with environmental laws and standards.
- Utilization of life cycle assessments methods and tools.
- Special educational and professional qualifications of staff.
- Existence of conformity certificates.
- Investments in research and sustainable development.
- Use of recognized eco-labels (4).
- Evidences from third-party agencies.
- Ability to apply environmental measures.

(1) In order to estimate the cost of green products the life cycle costing (LCC) is applied. LCC refers to all the costs that will be incurred during the lifetime of the product: acquisition costs, operating costs, maintenance costs and disposal costs.

(2) To give an example, most of the following must hold for the products to be procures:

- Their purchase must not only be purposeful, it must be necessary.
- They must decrease energy consumption.
- They must have a small or minimal environmental impact.
- They must have a small overall cost of ownership.
- They must be procured from local suppliers.
- Their packaging must be recyclable and others.

(3) In brief, environmental standards are tools supporting the continuous efforts by organizations to manage more effectively and efficiently their environmental impacts. The don't demonstrate the practical application of green practices but rather their commitment to the implementation of activities aiming to the continuous improvement of their environmental performance. The principal environmental standards are EMAS, applied by organizations active in European Union nations, and ISO 14001, which is a global standard.

(4) The placement of eco-labels on a product demonstrates – after strict checks, external verification and continuous tests – that a product satisfies specific and pre-designated environmental criteria. Environmental criteria take the entire life cycle of a product into consideration. Or, it simply demonstrates that it complies with specific technical specifications (as mentioned previously, the corresponding environmental criteria also cover the technical/environmental

specifications for the realization of procurements). Criteria are established and documented by international or national independent organizations (such as the International Organization for Standardization, the European Commission, US Environmental Protection Agency, etc.), while the process relating to the checking thereof is regularly applied by specific third-party authorities.

For example, according to the internet site http://ec.europa.eu/environment/ecolabel/ documents/light_bulbs.pdf, the conditions for the purchase of light bulbs are: (a) for energy efficiency: 10% better than the lumen per watt value according to Class A; (b) regarding the lifetime (hours): in case of single ended: 15,000 hours and in case of double ended: 20,000 hours; and (c) about mercury (in mg) must be less than 1.5 or 3.0, accordingly. It should be noted at this point that there are cases of products which while they meet the standards established by recognized eco-labels, their manufacturers have not affixed an eco-label on them or their packaging. This is due to the fact that the process for acquiring an ecological marking entails a high cost.

Readers are recommended to visit the internet site www.ecolabelindex.com/ecolabels/ which is, as the home page states, "the largest global directory of ecolabels, currently tracking 465 ecolabels in 199 countries, and 25 industry sectors" as well as the internet portal of Global Ecolabeling Networking (GEN): www.globalecolabelling.net/what-is-eco-labelling/ where the various types of eco-labels are presented and analyzed.

Additionally, visiting the internet sites of organizations managing eco-labels is also of interest; these include: http://ec.europa.eu/environment/ecolabel/index_en.htm, www.iema.net/ems/, http://ec.europa.eu/environment/emas/index_en.htm, ww.bsigroup.co.uk, www.biffa.co.uk/getrecycling/symbols.php, www.bafts.org.uk, www.pro-e.org, www.recyclenow.com, www.fairtrade.org.uk, www.rainforest-alliance.org, www.wwf.org.uk, www.wfto.org, www.redtractor.org.uk, www.leafuk.org/leafuk/, www.msc.org, www.rspca.org.uk, www.organicfarmers.org.uk, www.orgfoodfed.com, www.organic-trust.org, www.fsc.org, www.tcodevelopment.com, www.carbon-label.co.uk.

Moreover, according to the Article 57(4) (a, c, g and h) of Directive 2014/24/EU, exclusion criteria can be:

- Non-compliance with applicable international and/or national environmental laws.
- Grave professional misconduct which renders integrity questionable.

- Significant/persistent deficiencies in performance of substantive requirement under prior contract which led to termination or comparable sanctions.
- Misrepresentation of any of the above or inability to submit supporting documents.

> The previous criteria are frequently the subject of international and standardized documents, such as, for example, the European Single Procurement Document (ESPD) which functions as preliminary evidence of compliance with exclusion and selection criteria. According to the Article 59 of Directive 2014/24/EU, this is an updated self-declaration form, which exists in a standard electronic format, and allows suppliers to confirm their compliance with the grounds of exclusion and selection regarding the procurement of environmentally friendly products.

At this stage, the procurement manager is faced with various questions. For example, to procure or own-manufacture? To have a small or an extended supply base?; that is, to have one or many suppliers?

Thus, for example, for the green procurements in the case of a supplier and besides all other advantages and specifically securing better prices due to the purchase of larger quantities (discounts) and the establishment of long-term relations of interest are also lower transportation costs since larger quantities are bought and, consequently, a smaller number of routes take place. The following are some of the advantages to having many suppliers: better prices and terms for the contracts due to competition, increase of the suppliers' interest and technological development, and protection against strikes, natural disasters, etc.

Another issue for green procurements is the *selection* of a *local or international* supplier. It is true that when a company selects a supplier from the international market, it attains better prices, better quality and a greater range of products. If, indeed, the supplier uses green practices or/and material for the production, packaging and distribution of the products, then the result is satisfactory. Green procurements support the local and short supply chains models. Opting for local chains undoubtedly supports local producers and the products from a particular region. By opting for a local supplier, the business can maintain lower inventories, due to the short distances involved, attain reliability in deliveries, establish long-term and strategic relationships, satisfy extraordinary orders and attain lower prices due to the decreased transportation cost. For example, in the case of a food manufacturer, local food supply chains refer to all the logistics activities that are related with the production, packaging, storage, transportation, distribution and consumption of foods in a particular location. For food producers, local supply chains have become a new strategy in order to meet customers' environmental and ethical demands for products sourced closer to local areas.

A critical factor for the success of green procurement is the implementation of e-procurement practice, namely the performance of the green supply cycle electronically (hence the "e"), by utilizing internet practices and technologies. Modern internet use has automated the tasks of the supply cycle from the expression of interest all the way to the online filling out of the payment order and the recording of the products received. For example, the buyer may publicize a request for information (RFI), request for proposal (RFP) or request for quote (RFQ) on his own web site and/or send e-mails to specific suppliers or special e-marketplaces. The buyer can then identify potential suppliers from those who responded to the online request for bids/offers (usually an RFQ or RFP), while the latter may download the information regarding the required products. But an auction can also be performed digitally, by the submission of tenders inside a specific and usually short period. The buyer assesses the offers submitted by the suppliers, while all communication and negotiations are carried out electronically and the announcement of the award is also performed online. It is manifest that the principal objectives of e-procurement relate to the automation of the supply cycle, the improvement of the procurements process (with respect to both quality and time), the decrease in the usage of "paper" (uses fewer, if any, printed documents), the eradication of errors and, of course, the decrease of the overall cost.

The main question here regards which procedure will be employed for the assessment and ultimate selection of suppliers. A simple methodology for selecting the suitable supplier is the scoring model. The steps for the implementation of the methodology are the following:

1) Selection of the evaluation criteria (both inclusions and exclusions, as established earlier).
2) Establishing the weight of each criterion (based on the importance attributed to it in the drafting and concluding of the contracts – next stage).
3) Evaluation of each supplier with respect to every criterion using the scoring model.
4) Calculation of each supplier's scope for each criterion.
5) Calculation of the total/final score for each supplier.

Say, for example, that a company after evaluating the submitted offers has concluded in considering three prospective suppliers: A, B and C (Table 4.3).

The first column of the Table 4.3 presents the evaluation criteria selected by the company's executives, the second column the weight (i.e. importance) of each criterion, the third, fourth and fifth column the score on a 1–5 scale (where 1 is the lowest and 5 the highest) for each choice (A, B and C) while the last columns (Rank A, Rank B and Rank C, respectively) show the score

Table 4.3 Example of scoring model

Criteria (inclusions/exclusions)	*Weight*	A	B	C	*Rank A*	*Rank B*	*Rank C*
Cost of green products	30	4	3	4	120	90	120
Compliance with applicable international and national environmental laws	20	4	4	3	80	80	60
Compliance with international, national and business environmental standards	20	5	4	3	100	80	60
Experience in environmental products	10	4	3	2	40	30	20
Efficiencies in performance of substantive requirement under prior contract	10	4	4	5	40	40	50
Use of recognized eco-labels and/or evidences from third-party agencies	5	4	3	3	20	15	15
Utilization of life cycle assessments methods and tools	5	4	4	5	20	20	25
Total:	100				420	355	350

of each supplier for every criterion, which ensues if one multiplies the weight for each criterion with the corresponding score. By adding the products, a total score results: 420 for supplier A, 355 for supplier B and 350 for supplier C. The results show that Supplier A, who scored the highest, leads the process of evaluation and selection.

The methodology developed here is simple and gives to the executives of the procurements department and management, in general, the ability to study the process but also the real needs for green products and services. On the other hand, both the selection of the criteria as well as the assessment of their weight and the scoring thereof are characterized by subjectivity. This stage is completed by awarding the contract to the selected supplier.

Fifth stage: finalizing the contract. Technical – environmental specifications, as well as inclusions and exclusions – criteria constitute the reasons to be taken into account at the performance of the contract. The contract delimits the obligations to which the supplier must respond after the award of the contract. They do not simply relate to product specifications/characteristics but also relate to the parameters/functions of the logistics system, such as the transportation/distribution of the products, the selection of appropriate (environmentally friendly) transportation, the selection of the routes and delivery times, and the use of telematics applications to replace accounting and transportation printed documents. Experience has shown that there are many things not taken into account. For example, what would be the value of the procurement of green products when such products are transported by truck in major urban centers and at peak traffic hours, etc.

Sixth stage: measuring the contract performance. A critical factor for the success of the management of the supply cycle is measuring the degree of its success. It is for this reason that the general key performance indicators are applied, indicators such as quality of products and/or services, on-time delivery, compliance with contract and/or service level agreements, cost savings, procurement lead/cycle time, inventory turnover ratio, etc., as well as specific environmental indicators such as the reductions in of CO_2 or other greenhouse gas emissions, toxic materials, solid waste and energy use over the life cycles of the products increasing recycled content, the degree of the application of green practices, standardized packaging, etc. Perhaps the most critical performance measurement index is the degree for the decrease of environmental impacts with the simultaneous increase of the cost/performance ratio.

The management of these indicators is undertaken by the procurements department, which keeps historic date in a specific file wherein possible noncompliance of suppliers to the clauses in the contracts will also be kept. But, above everything else, what is called for is the enactment of systematic and quantified reviews of the green procurement program and the impact of their green procurement programs. Such a review may frequently bring about modifications in the green procurement program by seeking more environmentally friendly products and, of course, more reliable and cooperative suppliers. Indeed, it is for this reason that a third party may be contracted to monitor compliance of the supplier.

An alternative method for selecting and evaluating suppliers is the set pair analysis (SPA). Set pair analysis is a multi-criteria analysis method, in which the alternatives (suppliers) are examined and compared in pairs. The main advantage of set pair analysis as a multi-criteria assessment method is that both qualitative and quantitative selection criteria are taken into account, although high bias of qualitative criteria ranking can be also considered as a disadvantage, e.g. qualitative criteria weighting, or performance of alternatives based on qualitative criteria. To avoid this problem, the researcher should use the most objective methods possible (e.g. market research to select and evaluate the qualitative criteria).

The application steps of the set pair analysis are as follows:

1) Selection of assessment criteria.
2) Evaluating the selected criteria by the set pair assessment (weighing each criterion relatively to the others in pairwise comparison).
3) Scoring alternatives on each criterion.
4) Overall assessment of alternatives based on qualitative criteria.
5) Assessing alternatives based on quantitative criteria.
6) Final assessment of alternatives based on qualitative and quantitative criteria.

To better understand the set pair analysis, following is an example on selecting the purchase of green products using three alternative suppliers (I, II, III).

The purchase cost of green products for each supplier:

- Supplier I: €3.400
- Supplier II: €4.300
- Supplier III: €4.000

Step 1: selection of assessment criteria

The management of the company decided to use the following qualitative criteria:

A) Compliance with applicable international and national environmental laws and standards.
B) Experience in environmental products.
C) Efficiencies in performance of substantive requirement under prior contract.
D) Use of recognized eco-labels and/or evidences from third-party agencies.
E) Utilization of life cycle assessments methods and tools.

Step 2: evaluating the selected criteria

Initially a comparison of each criterion in relation to all the others is conducted. The comparison is executed pairwise (Table 4.4). The sum of a pair's score is equal to one (e.g. [score_A_in_relation_to_B] + [score_B_in_relation_to_A] = 1)

In Table 4.5, the global weightings of each criteria are calculated. First the values of each line are summed. The weighting of each criterion is the result of dividing the sum by the number of the pairs under evaluation.

From the results of Table 4.4, we conclude that criterion B is the most important with a weighting of 26%, followed by A with 24% etc. Criterion E is the least important with 8%.

Step 3: scoring alternatives for each criterion

In this step, the assessment of alternatives (choice of supplier) is carried out, evaluating every alternative by each qualitative criterion with the pair method (Tables 4.6–4.10). Once the alternatives have been rated for their performance on each criterion, the sum of each line is calculated. The performance of each alternative is computed by dividing the line sum by the number of pairs examined.

From the results of Table 4.6 we can conclude that alternative (supplier) I meets the qualitative criterion A (compliance with applicable international and national environmental laws and standards) better with a weighting of 40,00%, followed by III with 36,67%. Alternative II has the lowest performance with a weighting of 23,33%.

Table 4.4 Weighing qualitative criteria in pairs

	A	*B*	*C*	*D*	*E*
A	–	0.4	0.7	0.5	0.8
B	0.6	–	0.6	0.6	0.8
C	0.3	0.4	–	0.7	0.9
D	0.5	0.4	0.3	–	0.7
E	0.2	0.2	0.1	0.3	–

Table 4.5 Calculation of criteria weightings

	A	*B*	*C*	*D*	*E*	*Line Sum*	*Number of Pairs*	*Criterion Weighting*
A	–	0.4	0.7	0.5	0.8	2.4	10	24,00%
B	0.6	–	0.6	0.6	0.8	2.6	10	26,00%
C	0.3	0.4	–	0.7	0.9	2.3	10	23,00%
D	0.5	0.4	0.3	–	0.7	1.9	10	19,00%
E	0.2	0.2	0.1	0.3	–	0.8	10	8,00%
Total:								100,00%

Table 4.6 Assessing alternatives by criterion A

A. Compliance with applicable international and national environmental laws and standards						
Alternative	*I*	*II*	*III*	*Line Sum*	*Number of Pairs*	*Criterion Weighting*
I	–	0,7	0,5	1,2	3	40,00%
II	0,3	–	0,4	0,7	3	23,33%
III	0,5	0,6	–	1,1	3	36,67%
Total:						**100,00%**

Table 4.7 Assessing alternatives by criterion B

B. Experience in environmental products						
Alternative	*I*	*II*	*III*	*Line Sum*	*Number of Pairs*	*Criterion Weighting*
I	–	0,3	0,3	0,6	3	20,00%
II	0,7	–	0,7	1,4	3	46,67%
III	0,7	0,3	–	1	3	33,33%
Total:						**100,00%**

Table 4.8 Assessing alternatives by criterion C

C. Efficiencies in performance of substantive requirement under prior contract						
Alternative	*I*	*II*	*III*	*Line Sum*	*Number of Pairs*	*Criterion Weighting*
I	–	0,8	0,6	1,4	3	46,67%
II	0,2	–	0,6	0,8	3	26,67%
III	0,4	0,4	–	0,8	3	26,67%
Total:						**100,00%**

Table 4.9 Assessing alternatives by criterion D

D. Use of recognized eco-labels and/or evidences from third-party agencies						
Alternative	*I*	*II*	*III*	*Line Sum*	*Number of Pairs*	*Criterion Weighting*
I	–	0,8	0,2	1	3	33,33%
II	0,2	–	0,2	0,4	3	13,33%
III	0,8	0,8	–	1,6	3	53,33%
Total:						**100,00%**

Table 4.10 Assessing alternatives by criterion E

E. Utilization of Life Cycle Assessments methods and tools						
Alternative	*I*	*II*	*III*	*Line Sum*	*Number of Pairs*	*Criterion Weighting*
I	–	0,4	0,5	0,9	3	30,00%
II	0,6	–	0,6	1,2	3	40,00%
III	0,5	0,4	–	0,9	3	30,00%
Total:						**100,00%**

Based on the results of Table 4.7, we can infer that supplier II fulfills the qualitative criterion B (experience in environmental products) better with a 46,67% weighting. Supplier III is rated at 33,33% and the lowest performance by this criterion belongs to supplier I at 20,00%.

With regard to criterion C (efficiencies in performance of substantive requirement under prior contract), the assessment of alternative suppliers reveals that supplier I responds best at 46,67% and supplier I and II take second place with the same rating of 26,67% (Table 4.8).

Based on the results of Table 4.9, we conclude that supplier III meets the qualitative criterion D (use of recognized eco-labels and/or evidences from third-party agencies) best with a 53,33% score and followed by supplier I at 33,33%. The lowest performance by this criterion belongs to supplier II at 13,33%.

Regarding the criterion E (utilization of life cycle assessments methods and tools), the evaluation of alternative suppliers shows that supplier II performs best at 40,00% and suppliers I and II follow with an equal performance of 30,00% (Table 4.10).

Step 4: overall assessment of alternatives based on qualitative criteria

In this step, the final scoring of the alternatives, based on the qualitative criteria, takes place. The overall performance of each supplier relatively to the qualitative criteria is calculated from the score and the corresponding weight. The results of the overall assessment based on qualitative criteria is presented in Table 4.11.

In conclusion, the evaluation of suppliers based on qualitative criteria revealed supplier III as the optimal solution with a score of 36,13%. Second best choice would be supplier I at 34,26% and last supplier II at 29,60%.

Step 5: assessing alternatives based on quantitative criteria

In this step, the evaluation of alternatives is carried out based on quantitative criteria. The quantitative criteria are objective data that relate to purchase costs of green products per supplier. The collection of quantitative data can be accomplished by means of a market research. In Table 4.12,

Table 4.11 Overall assessment of alternatives based on qualitative criteria

Selection Criteria	*Criteria Weightings (1)*	*Alternative I*		*Alternative II*		*Alternative III*	
		Scoring on each Criterion (2)	*Weighted Score (1) × (2)*	*Scoring on each Criterion (3)*	*Weighted Score (1) × (3)*	*Scoring on each Criterion (4)*	*Weighted Score (1) × (4)*
A	0,24	0,40	0,096	0,233	0,05592	0,367	0,08808
B	0,26	0,20	0,052	0,467	0,12142	0,333	0,08658
C	0,23	0,467	0,10741	0,267	0,06141	0,267	0,06141
D	0,19	0,333	0,06327	0,133	0,02527	0,533	0,10127
E	0,08	0,30	0,024	0,40	0,032	0,30	0,024
		Sum	**0,34268**	**Sum**	**0,29602**	**Sum**	**0,36134**
		Or	**34,26%**	**Or**	**29,60%**	**Or**	**36,13%**

Table 4.12 Rating of alternatives based on quantitative criteria

	Alternative I	*Alternative II*	*Alternative III*
Purchase Cost (C)	3.400	4.300	4.000
$C^{-1} = 1/C$	0,000294118	0,000232558	0,000250000
$C x \pounds (C^{-1})$	2,6407	3,3397	3,1067
$\left[C x \pounds (C^{-1})\right]^{-1}$	0,3787	0,2994	0,3219
Rating/Percentage	37,87%	29,94%	32,19%

the sequence of actions aiming to evaluate/categorize suppliers based on the purchase cost of green products is shown.

Based on the results of Table 4.12, the solution of supplier I is qualified at 37,87%, followed by supplier III at 32,19% and the last choice would be the purchase of green products from supplier II (percentage 29,94%).

Step 6: final assessment of alternatives based on qualitative and quantitative criteria

The final step of the set pair assessment is the evaluation of alternatives based on both qualitative and quantitative criteria. For this reason, the researcher is called to assign weightings to qualitative and quantitative criteria. In this particular example, the management decided to assign the following weight factors to qualitative and quantitative criteria:

- 70% weighting on qualitative criteria.
- 30% weighting on quantitative criteria.

In Table 4.13, the final assessment results of the suppliers are depicted, upon which the final selection decision is based.

By analyzing the results of Table 4.13, it is shown that the best choice is Supplier I at 35,34%, while Supplier III follows with a slight difference at 34,95% and the last choice is Supplier II at 29,70%.

Barriers to broader adoption of green procurement

There are several barriers to the broader adoption of green procurement practices by organizations. These can be categorized in three groups: 1) product and market; 2) organizational and staff; and 3) information and tools. In more detail:

Product and market

- Perception that environmentally friendlier products and services may be more expensive than conventional ones. This is probably due to the fact that green products are usually not purchased in large quantities and, consequently, there are no scale economies and lower prices due to discounts. One must also not overlook the fact that purchasing costs may in many cases be higher, but other costs are probably lower. In other words, when products are examined from a life cycle perspective, green products are frequently cheaper. For example, transportation costs may be smaller, since less packaging is used while the disposal cost is also lower. Contributing to this is also the fact that life cycle costing is not easy to apply. However, one should also not overlook the fact that the benefits of green procurements (as were previously analyzed) are often higher than its purchase cost.
- Perception that environmentally friendlier products and services are not as mature as conventional ones, since the technologies involved in their manufacture are still at an experimental stage.
- Perception that environmentally friendlier products and services are "second-class", quality-wise as compared to conventional ones.
- Lack of a range and variety of green products in the market, especially local markets. This is attributed to the cautiousness of green product suppliers who recognize many risks in an uncertain and rapidly changing market.
- Lack of common/broadly accepted assessment and implementation environmental standards, as well as regulations.
- Lack of suppliers with environmental certifications and credentials, as well as suppliers that have eco-labels in their products and/or packaging.
- Lack of suppliers' involvement in the performance of environmental programs. Truth be told, the involvement of suppliers is detected principally in the procurement process. They could also be involved in the planning for environmental strategy and the improvement of environmental performance.
- Policy makers, principally on the national level, do not collaborate with the market and specifically the manufacturers and suppliers of

Table 4.13 Final assessment of alternatives

Selection Criteria	*Criteria weightings (1)*	*Alternative I*		*Alternative II*		*Alternative III*	
		Rating of alternative on each criterion (2)	*(1) × (2)*	*Rating of alternative on each criterion (3)*	*(1) × (3)*	*Rating of alternative on each criterion (4)*	*(1) × (4)*
Qualitative	0,7	0,3426	0,23982	0,296	0,2072	0,3613	0,25291
Quantitative	0,3	0,3787	0,11361	0,2994	0,08982	0,3219	0,09657
		Sum	0,3534	**Sum**	0,2970	**Sum**	0,3495
		Or	35,34%	Or	29,70%	Or	34,95%

environmentally friendly products, neither do they finance or, in general, encourage green procurement programs.

Organizational and staff

- Lack of integrating green procurement programs, which are aligned to the organizational strategy and business values.
- Lack of top management support and commitment in green procurement programs.
- Absence of a corporate policy statement outlining the corporate commitment to green procurement.
- Lack of knowledge about the critical issues of the environment, the concepts and terminology of green procurement and how to develop environmental criteria.
- Lack of training, culture and technical knowledge for procurement staff on environmental management standards in the green procurement cycle and specifically in the application of environmental criteria and techniques – environmental specifications. Moreover, there is also the reactionary attitude of personnel against any change. Many times, employees are not aware of the green procurement aims, objectives and policies.

Information and tools

- Lack of integrated information systems for managing (planning, execution, monitoring and control of) the supply cycle.
- Difficulty in collecting information (frequently due to the lack of updated, reliable and multi-lingual sources) for all stages of the supply cycle and specifically for market-research into green products and suppliers, as well as the corresponding regulatory and legal framework.
- Lack of information sharing among supply chain members.
- Lack of web application tools, as well as, checklists and handbooks.
- Absence of monitoring mechanisms (methods and tools) to evaluate the performance of green procurement programs and strategy.

All of these obstacles and challenges reinforce the view of many academics and practitioners that the best approach to achieve sustainable procurement is in an incremental manner. In simpler terms, this procedure is a continuous effort which may commence in the procurement of a small number of green products that will grow and will relate to the procurement of increasingly larger quantities of such products. Moreover, it must follow and observe specific principles: It must be a systematic/step-by-step process (as described in the section on the supply cycle), and it must take account of the modern market and society requirements for a sustainable policy, to bolster the local market, rest on a strict regulatory framework and employ specialized tools and approaches.

Chapter summary

Green procurement is a derivative of extending green practices throughout the supply chain. In accordance to other links in the supply chain undergoing green transformation, it is governed by both economic and environmental forces that shape decisions. In this chapter, the life cycle of procurement, as well as planning and common practices, were outlined and applied examples were presented. To formalize and ensure reliable and comparable performance, the labeling and certification schemes were presented. In order to describe green procurement without bias, this chapter also analyzed the barriers of broad adoption.

Notes

1 According to California Sustainability Alliance (2010) the key criteria for environmentally friendly/alternative green products are:

- Bio-based, biodegradable, compostable, (rapidly) renewable materials and recyclable, with recycled content and reduced packaging.
- Carcinogen-free, chlorofluorocarbon (CFC)-free, lead-free, less hazardous, low volatile organic compound (VOC) content, low toxicity, mercury-free and persistent bio-accumulative (PBT) toxics free.
- Durable, energy efficient, locally managed, reduced greenhouse gas emissions, refurbished, resource efficient, upgradeable and water efficient (CSA, 2010).

2 Results/lessons learned are in complete accord with the studies elaborated by independent third-party research organizations such as Aberdeen Group, ISM, AMR Research and Forrester.

5 Green production

Discussion questions

- What makes green production critical for a product's or service's environmental performance?
- How is design related to the environmental performance of a product or service?
- What are the stages of green production?

Introduction

Environmental protection and mitigation of impacts caused by human activity are issues that more and more companies around the world are trying to address. However, the purpose of any commercial activity is the creation of revenue from the satisfaction of consumer demand as well as the increase of the product's value. To that end, the integration of sustainable development into the value chain of a product is the new challenge that today's industry faces.

This turn towards sustainability creates an enormous potential for companies around the world.[16] Businesses are now reforming their practices by focusing on the improvement of not only the environment but also the society and the economy, integrating into their business plans these three pillars of sustainability. In order to implement such business plans, the contribution of all the involved in the products supply chain is essential. Thus, the main goal is to produce goods and provide services taking into account the environment, nature and people living in it, without causing any risk to their wellbeing.

This chapter focuses on the good practices and the steps that should be taken towards green production. Green production is an essential part of the green supply chain, especially considering that approximately 90% of the waste related to products are generated before the product reaches the consumer.[17] It is easily understood that the adoption of a life cycle approach in design and production is important in order to prevent environmental, social and economic impacts. The adoption of life cycle thinking requires a transformation of the existing practices regarding materials, technology and processes taking into consideration the environment and everything it entails.

The use of analytical methods for assessing environmental impacts during production of goods (as during their entire life cycle), is a key advantage for the development and the increase in the competitiveness of a business. By considering the environmental consequences of its operations, a business proves that it functions promoting transparency in its procedures and in its connections with its partners. All those involved in the production of a product (from the designer to the wholesaler) should recognize that the adoption of sustainable practices is an obligation of all. Therefore, the shift to green production requires the contribution of all stakeholders.

For their part, entrepreneurs should realize that improving the environmental performance of their business can lead to an overall improvement and bring plenty of benefits. The application of environmental thinking in green production can result in increase in revenue due to more efficient production processes and reduction of costs with the reasonable use of materials and resources. Also, the adoption of an environmental agenda in production has resulted in an increasing demand for consumer products. Consumers have a strong environmental awareness in recent years and require products that have been manufactured to specific standards and based on current legislation. Consequently, the use of standards (both quality and environmental) in production has led to a general increase in product quality.

The greening of production, along with the turn towards environmental protection and sustainable development, contributes to the modernization and development of businesses around the world. The scientific world and the industry have realized the importance of green production in promoting environmentally friendly and also cost-effective practices. To that end, many technologies, initiatives, instruments and products have been developed. Today, every business has access to a wealth of innovative environmental technologies and practices. Taking advantage of this opportunity, a business can have significant environmental benefits which in turn can satisfy consumers and lead to its rapid development.

Green production design

A large number of environmental impacts are linked to a product's life cycle. Also, the environmental consequences associated with a product depend directly on the type of product, the materials that compose the product and the way it is produced or manufactured. For example, the environmental impacts of making a plastic bottle vary from those associated with the production of a glass bottle. However, despite the intensity or the type of effects associated with a product, the design of its production is one of the most important stages of its life cycle.

When designing the production of a product, all the necessary decisions concerning its environmental behavior can be taken into account. As can be seen, the design of the production process for the manufacture of a product directly affects its overall environmental footprint. During production, the materials selected, using the technology chosen, are combined to produce the product. It is therefore perceived that the effects associated with the life cycle of a product are consolidated during manufacture. As a result, designing a production process that takes environmental impact into account is of the utmost importance.[18]

As mentioned earlier, when designing the production process, it is important to take into account environmental impacts. To achieve this goal, cooperation among all those involved in the production process is essential. Cooperation among all sectors of an enterprise can result in the optimization of the production process as all the individual criteria are considered. In this way, the environmental dimension can be integrated into the design of the production process, together with quality and cost, leading to green production.

The concept of green production, according to the United Nations, concerns production that respects the carrying capacity of the environment. Green production attempts to de-link economic growth and environmental degradation through improving efficiency in the use of resources and production processes. Green production promotes social and economic development and also reduces resource degradation, pollution and waste. Taking these factors into consideration, incorporating the contribution to the environment in the design of the production process, businesses can create a new revolutionary generation of products and services that are characterized by environmental friendliness. This can be helped by the contribution of many different specialties, as it gives the opportunity to consider all the aspects of the manufacturing of a product. The immediate result is the recognition and the addressing of the possible environmental impact at an early stage in the product's life cycle.

In order to assess the impact on the environment during product manufacturing, it is important to consider all the stages of the production process. By separating and categorizing the production stages, it is possible to identify the individual environmental weaknesses. This way individual environmental impacts can be addressed. The next section presents the stages that a company should follow in order to improve its environmental performance by turning towards green production.

Green production stages

Production processes worldwide abide by the relevant legislation that regulates issues concerning major environmental concerns like water and air pollution, waste management, toxic and hazardous waste management, global

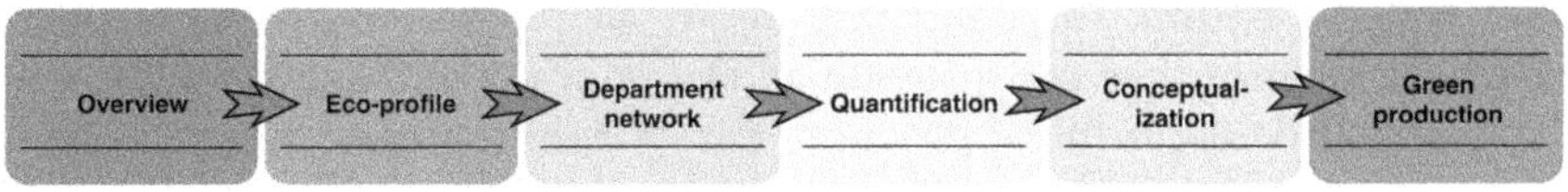

Figure 5.1 Stages of green production

warming, etc. Compliance with the legislation is a prerequisite for most of the companies that aim for satisfying their customers by providing them with high quality products. However, some companies take a step further by turning towards green production. The present section presents the process that a company should implement towards green production. The process is divided into six stages, as presented in Figure 5.1.

The first stage of a green production process is the overview stage. During the overview stage, the product's manufacture process is analyzed in order to identify the individual sub-processes that have environmental impacts. The second stage is the eco-profile stage, during which the environmental impacts of each individual process are categorized in order to facilitate their assessment. The third stage is the department-network stage. At this stage, a network of the different departments that contribute to the production process is created. By documenting the links between the different departments, it is easy to manage the information they provide in order to identify additional environmental impacts of the whole production process.

In the fifth stage, the quantification stage, the environmental impacts are quantified. The impacts are also evaluated against alternatives. After the impacts have been recognized, the conceptualization stage follows. During this stage, a strategy is formed in order to address the impacts that were identified during the previous stages. The final stage is the green production stage, during which the former assessment is implemented. In the next section, each stage is further analyzed.

Overview stage

The overview stage is probably the most important stage of the entire process. At this stage, an analysis of the production process and identification of all the individual stages is made. At the same time, the environmental impacts generated and the contribution of each stage during the production of the product is estimated. This creates a reference point on which the entire assessment is based. The distinction of the different stages in the production is essential to consider the various alternatives later in the process. In the first stage, a description of the flowchart of the production process is created. Also, all the possible impacts from the production are identified. The individual steps are recorded, together with the corresponding inputs (e.g. raw materials, energy) and outputs (e.g. waste, finished products).

During this first approach for the modeling of the production process and impact identification, the contributions of all the company's departments are essential since they can provide all the relevant information. The manufacturing process includes procurement of raw material or components, manufacturing (assembly included), packaging, transportation and distribution. In all these described stages, the respective inputs and outputs should be documented. It should be stated that all the individual processes should be recorded, regardless of the possibility or magnitude of the environmental impacts.

Eco-profile stage

At the previous stage, the stages and the environmental impact of the production process were recorded. During this stage, the data obtained from the previous stage are organized and categorized in order to create the ecological profile of the company. At the same time, an initial assessment of the possible origin of the impacts is made at this stage. The aim is to identify possible connections between the impacts at the individual stages of production. In this way, the stages of production can be arranged based on their environmental impact assessment. The environmental consequences can be classified into four groups of impacts depending on their relevance with materials, energy, chemicals and miscellaneous.

For impacts related to materials, their origin is examined as well as their dependence on limited resources. The energy aspect entails, among other components, all the energy types and resources that are used for production. Related to chemicals are all the impacts that derive from the use and disposal of chemical and hazardous substances throughout the production process. The miscellaneous group entails all the impacts that are not classified into the previous categories. The categorization of the impacts and their classification into groups can be presented in many forms to facilitate their interpretation. Indicatively, the matrix and the cluster diagrams are discussed.

Department-network stage (stakeholder analysis)

The majority of businesses in their shift towards green production start aiming at improving the environmental performance of their production processes. The initial stages, as presented previously, are very basic steps in managing the impacts of the production process as they form the basis upon which all subsequent improvements are made. However, this method also has some weak points, as with more complete recording and data collection, there is a considerable chance that a series of assumptions may be needed. The more assumptions, the greater the uncertainty in the final results.

In order to deal with the abuses mentioned, the individual departments of the business must be identified and recorded.[19] For each department, its role in the production process must be defined. Also, any connections/

overlays that the various departments have in the production process should be explored. This entire process can be defined as the department-network design. The design of the department-network can provide valuable information on the environmental impacts that result from the production process. It demonstrates the connections between the different departments and their contribution to the entire process, and as a result facilitates the impact identification process of the previous stages.

Quantification stage

In the product development process, a significant part of content generation is based on knowledge, experience and aptitude of the development team. Based on these assets, decisions and assumptions are made, even though hard data to support them is not available. Still, some of the choices need quantitative data in order to be evaluated correctly, as is the case with specifications and design choices that influence environmental impact. Assessments and methods estimating impact are based on numbers and recorded reliable data.

To engage in a green production procedure, a quantified life cycle assessment is executed, and its results will be also internally used in concept selection phases. A visualized and scaled representation of the environmental impact will reveal the order of magnitude of each manufacturing process, material or activity selection.

Life cycle assessment as part of the environmental assessment can be a tedious task, and the effort needed to complete the assessment varies greatly depending on the method implemented. Some simplified methods exist, and their selection depends mostly on the implementation team expertise. Also, the availability of concept-specific data can determine the suitability of methods, e.g. to what extend the materials used are both selected and dimensioned. The desired depth of analysis designates the methods complexity, whether it will be a simple pocket calculator or a computer tool. In Table 5.1 some of these methods are listed as reference for practitioners.

Another factor that production planners need to take into account is that outcomes of these methods aren't identical between them and are dependent on various factors. Obviously, the choice of examination model plays

Table 5.1 Simplified life cycle assessment methods

Method	*Country of origin*	*Material available*
Life Cycle Check	Denmark	https://goo.gl/QtkKHi
Ecodesignguide	Denmark	https://goo.gl/EK6bYu
Ecodesign PILOT	Austria	https://goo.gl/s843pj
Eco-Indicator 99	Netherlands	https://goo.gl/VMCFfY
ECO-it (Merged in SimaPro)	Netherlands	https://goo.gl/m8mxfb
EPS	Sweden	https://goo.gl/r7eHXP

a significant factor. Besides the internal processes of the tool chosen, the inputs to each model may vary because of tool design or because of assumptions defined by the user. For example, one might choose to include a manufacturing process or not. Data origin is also a major defining factor as primary data, or secondary data might be used. It is often the case that data might be imported from general models and not be process specific. Collection methods also influence data quality, or even data integrity. Last, the method's structured approach to the system might innately be inclined towards a specific area of focus and leave other aspects of the system in the background.

Assessing the environmental impact of complex production systems is a multifaceted problem that includes assumptions and selective data inclusion in order to bring order in an endlessly overlapping open-ended system. Therefore, one should always keep in mind that even full-fledged methods deliver results within a rather wide threshold of acceptance – and using quick shortcut methods like those presented in this chapter expands this threshold even further. The results are considered to be indicative and to be used as rough guidelines for early process design steps. Having followed the preceding steps of production design, most of the data inputs should be available and most of the production methods already defined. Teams that used such methods develop experience and, if the need is present, fine-tune databases reaching even higher levels of accuracy.

Conceptualization stage

Parallel to the traditional conceptualization of products and their production systems, in order to adopt a greener production system, environmental dimensions need to be considered. These constitute a complete environmental concept that embraces the product according to the information that was gathered and generated during the previous stages. Concept development is an open and creative synthesis process with numerous methods and tools in use, depending on the specifics of the desired outcome and the capacity of the team. Indicatively, some popular tools include semi-structured methods such as checklists, brain-writing and brainstorming, all in within the scope of environmental thinking. Less structured and freer are methods like "tomorrow headlines" and other future focused tools, whereby the impact of existing designs are evaluated or non-existent plausible solutions are presented in order to foster breakthrough thinking. Through the environmental mindset, this might lead to identifying weak spots or opportunities and initiate a creative and innovative adaptation to current data and capacity.

A structured method that can significantly aid the conceptualization stage is to use eco-design principles to promote environmental thinking. Following the ten principles of eco-design (Figure 5.2) leads to substantial improvement in environmental performance.

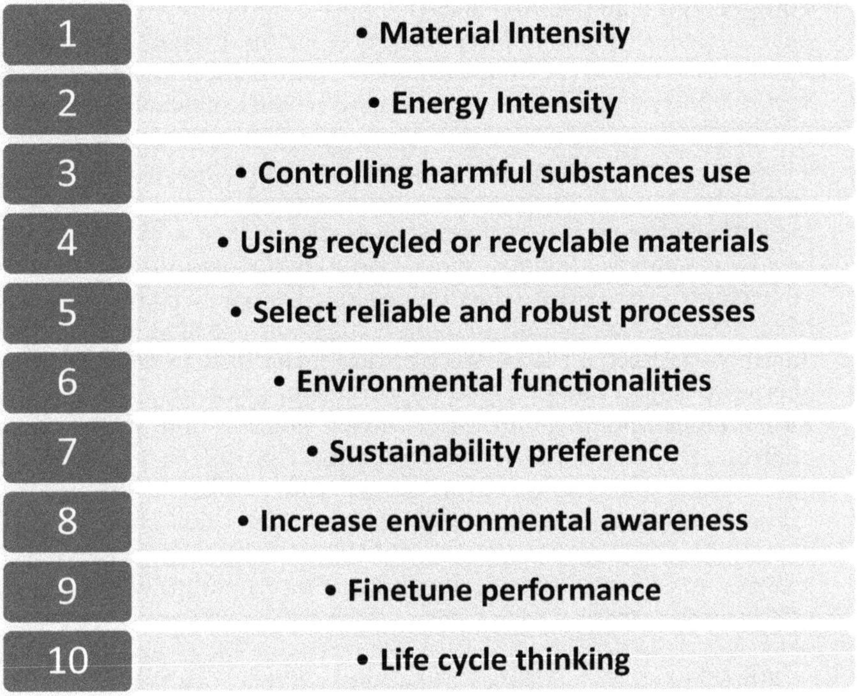

Figure 5.2 The ten principles of eco-design

- *Material intensity* is a factor that strongly influences all environmental variables. This is easily understood as the effects span covers the raw material extraction phase to the waste management and all processes in between, e.g. transportation costs related to handling and storing.
- *Energy intensity* can also account for a significant part of the environmental performance. As energy sources move from fossil-based to renewable sources, this factor will exert less pressure on the system. However, presently a friendlier energy resource usage is recommended.
- *Controlling harmful substances use* in the production process. For example, metal-cutting fluid emulsions need to be removed from chips to be separately handled and, if possible, alternative cooling and cutting agents explored.
- *Using recycled or recyclable materials* significantly increases the recovery rate of materials and utilizes streams of other products' waste management.
- *Selecting reliable and robust processes* that minimize the need for machinery replacement and ensure a high rate of reliability and therefore

increased running times and less overhead and fixed costs in all dimensions (energy, materials, financial).[20]

- ***Environmental functionalities*** integrated in the production line, e.g. standby of machines, heating and cooling optimization, energy recovery.
- ***Sustainability preference*** over alternatives can contribute to a noteworthy increase in environmental performance. Using sustainable alternatives at every decision point incorporates environmental benefits of these subsystems in the green production process.
- ***Increase environmental awareness*** even among production team members by communicating at every opportunity the superior environmental performance of the production process. This could include anything from detailed information to simple labels on machinery. This has a multiplying effect, as increased awareness can lead to feedback from alternative sources on potential improvements or caveats.
- ***Fine-tune performance*** of the production process. Even if the scheme implemented isn't optimal to reduce impact on the environment, the optimization of the process in service will offer a relative improvement.
- ***Life cycle thinking*** is to be considered as a first priority. Therefore, production planning can take place only after the life cycle of the process was examined first. Life cycle design can offer an outline of the production process design.

The aforementioned guidelines constitute a framework that in optimal conditions produces ideal production processes. Reality often compromises the feasibility of such concepts, even though all concepts include interesting and feasible features that should be identified and synthesized to define a green production scheme.

Green production stage

Successful completion of the previous stages leads to the last stage of compiling the district improvements into a universal strategy governing the green production process. Solutions and ideas need to be reviewed as part of a single entity, the green production scheme. Combining these parts in a comprehensive strategy should be done, keeping in mind that this plan needs to be applicable at all levels of implementation from upper management to the workforce. Thus, the team entrusted with this work has to consolidate and balance the needs of all groups along with conforming to green commitment. The work can be chaotic, so it is advisable to define a number of areas regarding the production plan and cluster solutions and ideas referring to them. As the proposals are being debated, they can be ranked and even if they don't present a short-term opportunity, they could have future potential and therefore should not be discarded. Records of these approaches can fuel future improvement attempts.

Effort and resources invested in this procedure are going to be translated into added value if the actual findings and results are implemented. An environmental improvement plan has higher adoption probability if it is aligned with the organization's strategic goals. To ensure seamless integration to the company's production process, the team needs to organize and present the agreed conclusions to transition to green production. The members of the team have to facilitate communication among the company's departments and instill the green practices and mentality to all participants.

Besides green production benefits, a number of positive side effects are realized in various areas:

- Corporate responsibility compliance and positive results to be publicized.
- Preparatory steps already taken for future environmental policies obligations.
- More detailed view of the production process life cycle.
- Innovation potential for voluntary environmental actions.

Chapter summary

Green production is an essential part of the green supply chain, especially if we consider that the vast majority of product waste is generated before the product reaches the consumer. It is easily understood that the adoption of a life cycle approach in design and production is important in order to prevent environmental, social and economic impacts. The adoption of life cycle thinking requires a transformation of the existing practices regarding materials, technology and processes taking into consideration the environment and everything it entails. In this chapter, the importance of green production design was highlighted, and the green production stages were analytically presented. Moreover, good practices and the steps that should be taken towards green production were discussed.

6 Outbound logistics (green transportation and distribution)

Discussion questions

- What is "green" transportation, and how is it differentiated from traditional practices?
- What are the factors that contribute to the "greening" of transportation?
- What are the key practices in transportation distribution?
- What are the guidelines related to "green" transportation?
- What are the expected benefits and challenges of green transportation?

Introduction

In recent years, there has been a shift in interest from the effect of environmental conditions on transportation execution (e.g. how weather conditions affect sea transportation) to the effect of transportation on the environment (e.g. how road transportation affects the increase of emissions of pollutants and other gases into the environment). Also, a trend from faster – but simultaneously, more polluting – means of transport to slower and more environmentally friendly means is observed.

Defining green transportation

"Green" transportation is the application of practices and technologies that aim to mitigate transportation's negative environmental impacts. "Green" transportation is part of an international economic practice which supports the implementation of an energy-effective and efficient system that will support a country's economic development (sustainable development) and the implementation of strict environmental standards that not only will not prevent entrepreneurship but on the contrary, will result in new innovative business initiatives (sustainability).

"Green" transportations are a part of "green" practices managing the logistics procedures (green logistics) that focus on:

- Designing and developing products adopting environmentally friendly practices and materials (like products manufactured from recycled materials, organic ingredients, products that use environmentally friendly packaging, etc.).[21]
- Selecting and cooperating with suppliers who also adopt environmentally friendly production and handling processes.
- Operating and using storage facilities that adopt energy saving mechanisms and technologies.
- Using recyclable packaging materials that do not pollute the environment.
- The efficient management of returned products (recycling of waste, etc.).

"Green" transportation initiatives are primarily aimed at changing the attitude of businesses and organizations towards major environmental issues like climate change and environmental degradation due to the inexorable use of natural and energy resources. To do this, however, their administrations should understand that:[22]

- The largest share of total energy consumption comes from transport (about 40%, according to the annual reports of the last two years of the European Union). From this, road transport consumes most of the energy (greater than 80%) (Figure 6.1).

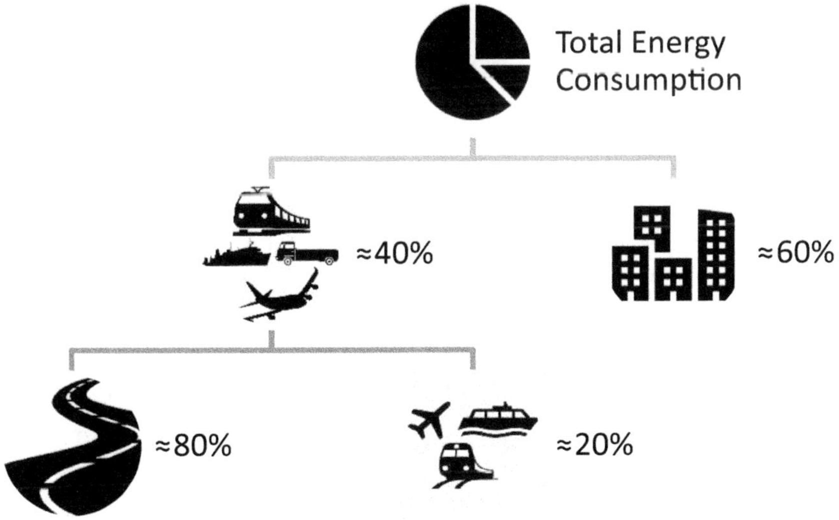

Figure 6.1 Allocation of energy consumption

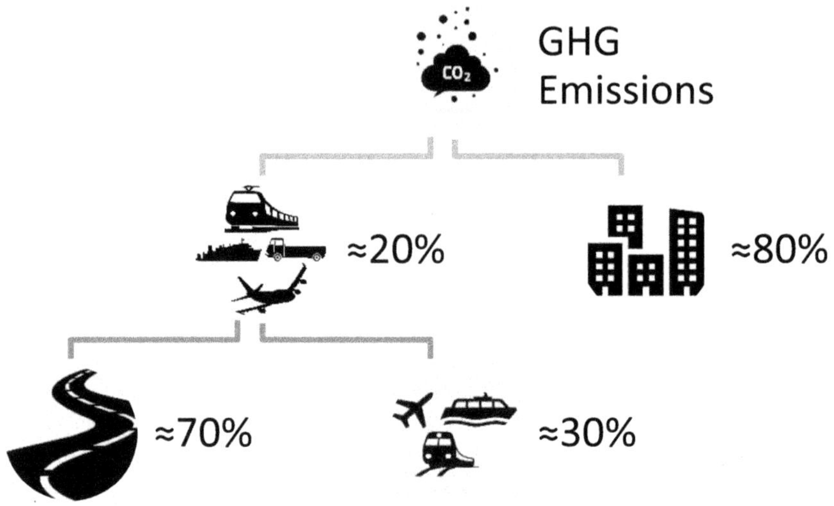

Figure 6.2 Allocation of GHG emissions

- 20% of emissions of greenhouse gases (GHGs) (such as carbon dioxide CO_2 but also other gases such as, CH_4, N_2O, SO_2, etc.) derive from transport, 70% of that from road transport (Figure 6.2).

Subsequently, these initiatives aim at identifying functions and daily tasks that directly or indirectly burden the environment and/or consume large amounts of energy. Finally, these initiatives aim at finding ways and mechanisms of reducing these concerns, while at the same time achieving economic and social goals. Consequently, the main objective of green transport is to implement a strategy whereby the economic objectives of an enterprise or organization must be linked to as little environmental impact as possible.

An example of setting targets is the European transportation policy, which proposes a number of initiatives grouped in four axes ([COM, 2011] 144 final / 28 March 2011), available at: http://eur-lex.europa.eu/LexUriServ/LexUriServ.do?uri=COM:2011:0144:FIN:en:PDF

1) Existence of "sustainability" in all transport sectors
2) Reduction of transportation's greenhouse gas emissions by 20% until 2020 and by 60% in 2050, compared to 1990.
3) Maintaining competitiveness at all levels.
4) Maintain the current level of mobility in every way.

Later in this chapter, the ten goals of the European transportation policy as they were introduced by the European Commission (2011) White Paper are described.

Green practices in transportation/distribution

Based on the previously mentioned factors, enterprises are led to:

- ***The preference and use of non-polluting means of transport.*** Such means of transport are the train and the ship or the combination of the either with the truck. Indeed, the combination of rail and sea transport with inland transport is the mode of transport proposed and promoted by the European Union for freight transport. Of course, the proposed means of transport do not have the speed, flexibility and reliability as well as the coverage offered by road transport, but on the other hand, the latter consumes the most energy and is the largest source of gas emissions.
- ***The use of an energy efficient transportation and distribution fleet,*** especially, on the use of trucks (means of transport in general) that consume clean energy (for example, there are hybrid systems that consume natural gas or biofuels, which can significantly reduce pollutants such as particulate matters [PMs], oxides of nitrogen [NO_x], volatile organic compounds[VOCs], unburnt hydrocarbons, benzine, etc.) or employ technologies that reduce fuel consumption (e.g. electric or hybrid vehicles with improved aerodynamic characteristics that can reduce fuel consumption from 10% to 20%.[23]). Also, the application of vehicle maintenance practices that ensures the reduction of energy consumed. Supportively, motives (e.g. tax relief) should be given for the replacement of old and heavy vehicles (>3.5 tons) with alternative vehicles or with new technology conventional vehicles.

> For example, in Europe, more than the 80% of used oils, is converted into biodiesel. In addition, about 30% of the fuels used by trucks for the transportation and distribution of McDonald's supply chain products come from biodiesel, while 40% of these fuels is produced from used cooking oils.[24–26]

- ***The design of efficient distribution networks*** in choosing locations for distribution centers or central warehouses near the points (markets) where the greatest demand is. This option achieves better and faster customer service, distribution routes become shorter and consequently the corresponding emissions of pollutant gases are reduced.[27,28] This practice is called near-shoring and is the reverse of off-shoring, which describes the production process in distant locations. Trains could also be selected for the transportation of goods/finished products from production points to distribution centers or central warehouses.

An example of an efficient distribution network design is the short supply chain that concerns products produced and consumed within a particular geographic area. Usually, a short supply chain is organized for supporting of a local food system. This system is a system of producing, processing and selling food within a defined geographical area. Examples of such systems are local farmers' markets, the sale of agricultural products by producers directly to consumers, restaurants, catering companies, retailers and others (farm-gate sales) and the community supported agriculture and public procurement schemes, where the source of food is located in a defined geographic radius. Foods traded within local systems are usually those for which traceability exists to their particular place of origin and have particular qualitative or other characteristics. Many times, this food is not processed or is slightly processed. To date, there is no official, commonly accepted definition of local food, nor is it a geographical definition of the term "local". The "local" character of these foods is usually empirical, compared to larger geographic regions, on a regional, national or global scale. Therefore, the beginning and end of a local area are subjective and depend on the content and purpose of each individual analysis.

The impacts (benefits) of the short supply chain can be distinguished in: (a) social; (b) economic; and (c) environmental impacts. In particular, the main social impacts of local food supply chains include: The development of social interaction and relations of trust between producers and consumers, the reinforcement of community feeling, the increase of consumer knowledge and the enhancement of understanding – from their side – of foods, agriculture and the environmental issues, which in some cases can even lead to the change of behavior.

The economic impacts of local food supply chains usually relate to agricultural development, the increase in local sales, the reinforcement of employment in local level, the increase in demand for the regional tourist products and the economic rebirth in a local level in general.

Environmental impacts usually are divided into two major categories. The first includes the energy use and the carbon footprint, while the second includes all the other environmental impacts. In this context, in the case of energy use and carbon footprint, the two popular methodologies for analyzing a chain are:

- Food miles, which is the distance traveled by the food from its point of production to its point of consumption and aims at the comparison of alternative chains, means of transport and food on the carbon emissions side, especially in transport.
- Life cycle analysis of a foodstuff, which seeks to assess the environmental impact of a product throughout its life cycle.

Generally, researchers' opinions vary regarding the magnitude of the benefits to the environment through the development and the dissemination of local supply chains. A large number of researchers believe that local supply chains can have significant positive environmental impacts, regarding carbon emissions and energy efficiency. On the other side, there are researchers who believe that local supply food chains and the concentration of economic activity in a place it is not certain to bring substantial environmental benefits, mainly because of small production volumes. Also, local food supply chains are believed to help reduce, or even eliminate, the use of pesticides and veterinary medicine, reduce soil and water contamination, reduce degradation of soil quality, enhance biodiversity, preserve water reserves, minimize food processing and minimize non-local inputs.

Generally, the environmental impact minimization, in the context of a supply chain, may be achieved when local supply chains have the following characteristics: They are local, seasonal and use environmentally friendly production methods.

Based on these findings, the following table lists three types of indicators that could potentially be used to assess the performance of a local food supply chain, as well as the related impacts associated with each sub-category of indicators.

Table 6.1 Basic types of environmental indicators and benefits in the context of local food supply chains

Potential indicator	*Potential environmental benefits related to the indicator*	*Issues to be discussed*
Local	• Reduced, related to transportation, greenhouse gas emissions.	Determination of "local". Probably, a definition for all European Union countries is not possible.

(*Continued*)

Table 6.1 (Continued)

Potential indicator	*Potential environmental benefits related to the indicator*	*Issues to be discussed*
Seasonal	• Reduced, related to storage, greenhouse gas emissions.	Seasons can be expanded, as for example with the use of heated greenhouses.
Ecologically friendly production methods	• Reduced, related to production, greenhouse gas emissions. • Reduced or no pesticide use. • Reduced soil and water pollution. • Reduced soil degradation. • Enhanced biodiversity. • Water preservation. • Minimization of processing: Reduction of greenhouse gases related to food production and storage. • Minimization of non-local inputs.	There is a wide variety in the selection of the cultivation system used by the producers. Should environmentally friendly means meet the majority, or part, of these benefits? Should suppliers achieve a part of these benefits to be considered environmentally friendly? Should the criteria be weighted, or even ranked? Is the reduction of non-local inputs feasible, or not so much due to the uneven distribution of resources?

- ***Better route design*** in order to achieve the right execution of merchandise transportation and distribution with the smallest number of routes and the reduction of redundant distances driven by trucks. IT and telecommunication applications (telematics applications) can make a significant contribution to this goal. For example, the knowledge of traffic on major road arteries may significantly reduce congestion, while better vehicle routing can reduce overall transportation costs, fuel costs and greenhouse gas emissions.[29]
- ***The recycling of old vehicles*** contributing to the reduction of environmental impacts (possible disposal of materials like catalytic converters, batteries, lubricants, etc., in landfills) on the one hand, with the removal of the reusable materials and on the other by their re-use as a source of energy. The recyclable parts of the trucks are engine, accumulator, sheet metal, trunk, aluminum wheels, copper, plastics, catalytic converters, refrigeration units and tires.[30,31]
- ***The increase of utilization degree*** in particular of trucks and the achievement of full truck load. Generally, the unifications of freight

transportation and distribution (groupage) allows the transportation of more goods with one means of transport. A common practice to achieve this is to define specific predetermined days during which trucks will serve specific areas. In this case, customers are informed about the route schedule and place their orders on this basis. In general, telecommunication systems facilitate the cooperation among businesses for the increased utilization of the capacity of their trucks, achieving scale economy which leads to reductions of costs and environmental impacts.
- ***Designing green-sustainable distribution networks***, primarily with the selection of the position of production units and distribution centers, according to the emissions of the reference system. A distribution center that is closer to the points of greater demand reduces the distance of the required routes. A very interesting practice is the introduction of the urban logistics consolidation centers.

Since the majority of the population lives in large urban centers, the transportation of goods in these centers represent 20% of the total road trips and causes 30% of air and noise pollution, the search for new practices in freight transport is imperative.[32] it has been measured, for example, that heavy-load transport at peak times in an urban center can increase fuel consumption by up to 100%.

The above-mentioned challenges and simultaneously, the existing restrictions (such as the failure to create and/or extend the urban road network, the existence of industrial zones and the concentration of a large number of private cars), demand the regional development through investment attraction, the exploitation of geographic areas near the urban centers and the support of combined transportation with the use of port terminals, airports, etc.

An answer to the foregoing challenges are the urban logistics consolidation centers. The implementation of an urban logistics consolidation center aims at increasing the efficiency of logistics procedures and the mitigation of their negative impacts in a large urban center, while at the same time aims at its sustainable development.

The benefits of the implementation of such a center are functional, economic, environmental and social. The concentration of logistic services and their provision by an experienced operator, who has the equipment and the appropriate infrastructure, will lead to a better and more efficient supply chain assessment in total and, of course, to the best customer service of the urban center. Also, it will contribute to the design and the provision of added value services. The economic benefits will come from scale economies that will be achieved by the provision and use of common infrastructures and means for the service of a large number of customers who are concentrated in the wider market. The environmental benefits are also expected to be significant because of the reduction of the number of routes in the city, the reduction of traffic congestion and damage to the environment from the use of this eco-friendly means of transport. Finally, the social benefits relate to the creation of new jobs in many specialties and in general the strengthening of the urban economy.

- ***The adoption and implementation of telematics systems and technologies.*** The logistics industry focuses on transport/distribution of products and, in particular, in the management of vehicle fleets. Placed on the market today are many applications and systems – intelligent transportation systems (ITs) – enabling real-time automatic vehicle location tracking, tracing problems on the vehicle or to any of its systems, vehicle navigation features, safety of drivers and vehicles, and monitoring of the status of the load.

  The use of telematics systems enables companies' logistics developers who use their fleet either to serve their own needs or the needs of others to control the activity of vehicles, drivers and goods from their computer screen at any time and from any place. Main elements in fleet management are vehicle (or fleet) detection in real time, recording route traces, viewing vehicles and routes on a map and releasing reports about the vehicle and its routes, allowing businesses to improve productivity, reduce their operating costs and abide by existing legislation.

  Navigation is usually a function provided by fleet management information systems which provide with a general overview of the company's means, routes and orders. Usually, these systems enable direct communication with drivers while cooperating with the rest of the enterprise's information systems, in order to monitor and check the order execution process. This ensures order pricing and the identification of the goods carried.

  A crucial point is the choice of the optimal route, whereby the transport or distribution of goods must be successfully completed in the shortest possible time, taking into account factors such as the volume, weight and value of goods, special requirements of goods in storage and distribution issues, the special requirements of customers regarding delivery times, existing time and geographic limitations, etc.

  For example, on many highways, an integrated telematics system operates that relays information for events such as congestion, accidents, etc. In particular, the system includes sensors providing information about the number, speed flow and density of vehicles, closed circuit television cameras that transmit the full picture of traffic conditions on the highway control center's screens, weather stations that are located in various places on the highway and provide with information

Green Practices in Transportation Distribution	*Preference and use of non-polluting means of transport*
	Use of an energy efficient transportation and distribution fleet
	Design of efficient distribution networks
	Better route design
	Recycling of old vehicles
	Increase of utilization degree
	Designing green -sustainable distribution networks
	Adoption and implementation of telematics systems and technologies
	Use of appropriate packaging
	Adoption of eco-driving habits

Figure 6.3 Green practices in transportation/distribution

on changes in the weather conditions, and noise and air pollution measurement stations.

In addition to the described information collection systems, others exist such as variable message signs along the highway as well as on its access ramps and toll stations, lane control signs, variable speed limit signs and exceeded permissible height detection devices of vehicles at the entrances. The benefits deriving from this navigation system's application are: (a) a more rational use of available resources (resources and personnel); and (b) reduction in response time of customers' orders and higher reliability in delivery times in general. These lead to an increase of the service level provided and productivity.

- ***The use of appropriate packaging*** such the use of thinner and lighter packaging, with the reuse and recycling of packaging materials.
- ***The adoption of eco-driving habits*** by all, and especially from the staff of the third-party logistics (3PLs) that may be supported by companies' administrations and also provide drivers with training programs about the ways of eco-driving for fuel saving. For example, drivers are trained in conservative driving and applying simple practices such as switching off the engine in small stops, the constant check of tire pressure and instant fix of various mechanic problems.

These practices may be combined with corresponding practices that can be implemented by passengers, particularly in urban areas. First, a change in behavior, lifestyle and movement with respect to the environment. Such examples are gait, use of bicycles, use of small capacity electric vehicles while at the same time reduce the use of private cars, sharing, etc. This of course requires the development of the appropriate infrastructure (as for example the creation of new bus lanes, the application of intelligent signaling systems, the greater use of fixed track transport means, the creation of walkways and bicycle lanes, etc.) and of course the change of the population's mentality (with the provision of training programs, the driver and instructors training and the use of informative advertising campaigns). It also requires an efficient management of urban mobility by implementing appropriate urban mobility actions in foodservice places such as parks, sports venues and shopping centers, hospitals, other public spaces, etc. A graphical summary of the above is presented in Figure 6.3.

European Union sustainability guidelines

The ten objectives of the European transport policy[33] are presented in this section as an informed framework employing green transportation practices while complying with government decrees. These are divided in three major groups.

Developing and deploying new and sustainable fuels and propulsion systems:

1) Halve the use of "conventionally-fueled" cars in urban transport by 2030; phase them out in cities by 2050; achieve essentially CO2-free city logistics in major urban centers by 2030.
2) Low-carbon sustainable fuels in aviation to reach 40% by 2050; also, by 2050 reduce EU CO_2 emissions from maritime bunker fuels by 40% (if feasible, 50%).

Optimizing the performance of multimodal logistic chains, including by making greater use of more energy efficient modes:

3) 30% of road freight to be transported more than 300 km should shift to other modes such as rail or waterborne transport by 2030, and more than 50% by 2050, facilitated by efficient and green freight corridors. Meeting this objective will also require appropriate infrastructure to be developed.
4) By 2050, complete a European high-speed rail network. Triple the length of the existing high-speed rail network by 2030 and maintain a dense railway network in all member states. By 2050, the majority of medium-distance passenger transport should go by rail.
5) A fully functional and EU-wide multimodal TEN-T "core network" by 2030, with a high-quality and -capacity network by 2050 and a corresponding set of information services.
6) By 2050, connect all core network airports to the rail network, preferably high-speed; ensure that all core seaports are sufficiently connected to the rail freight and, where possible, inland waterway system.

Increasing the efficiency of transport and of infrastructure use with information systems and market-based incentives:

7) Deployment of the modernized air traffic management infrastructure (SESAR [12]) in Europe by 2020 and completion of the European Common Aviation Area. Deployment of equivalent land and waterborne transport management systems (ERTMS, ITS, SSN and LRIT, RIS). Deployment of the European Global Navigation Satellite System (Galileo).
8) By 2020, establish the framework for a European multimodal transport information, management and payment system.
9) By 2050, move close to zero fatalities in road transport. In line with this objective, the EU aims at halving road casualties by 2020. Make sure that the EU is a world leader in safety and security of transport in all modes of transport.
10) Move towards full application of "user pays" and "polluter pays" principles and private sector engagement to eliminate distortions, including harmful subsidies, generate revenues and ensure financing for future transport investments.

Expected benefits and challenges of green transportation/distribution

The implementation of these practices will primarily benefit the living standards of people in the application area by reducing pollution and improving the living environment and their everyday life.

Greater benefits will also be gained by businesses that apply "green" practices. Initially, they will significantly improve their image and reputation in the market by enhancing their social responsibility actions. By creating "green" products and handling them with "green" transports, businesses will attract consumers and collaborate with other businesses with environmental consciousness and worries. They will also reduce their operating costs by reducing the energy and natural resources they consume to produce their products and services, while with reusing raw materials and packaging materials will also reduce production and packaging costs.

There are three main factors for the successful implementation of the practices mentioned in the previous section:

1) Developing an increased environmental awareness of consumers who require products that are produced and traded by companies that adopt "green" practices.
2) Turning businesses into adopting "green" practices aimed at reducing their operating costs, providing services and offering environmentally friendly products.

3) The introduction of rigorous policies (at national, European and global levels) and incentives (for example, replacement of old-technology trucks) by the state for enterprises to implement these "green" practices as well as inform customers (citizens) in order to increase their environmental awareness.

Chapter summary

"Green" transportation focuses on the reduction of the environmental impacts of commercial transportation. The main goal is to reduce the energy consumed, as well as the emissions of pollutants and gases that burden the environment. These efforts are aided by technology on one hand, and the strict legal framework governing transport, and particularly road transport, on the other. The benefits of "green" transportation in the long run exceed any investment costs for the application of "green" practices. In this chapter, green transportation and distribution was defined and analyzed and its important role in the supply chain highlighted. The European Union sustainability guidelines were presented, and the expected benefits and challenges of green transportation were discussed.

7 Green packaging

Discussion questions

- What is the role of packaging in production?
- What is an environmental label, and how are labels categorized?
- What are the most common examples of environmental labeling?
- What are the most essential features that a reliable eco-labeling system should have?
- What are the advantages and disadvantages of the use of environmental labels?
- How are environmental labels used in different functional areas of production?

Introduction

Packaging has played a very important role in everyday life. Materials and techniques have changed over the years, affecting quality of life. Today, very few products exist that don't need some form of packaging.

The term "packaging" refers to all the materials used for the protection, the assessment, the transportation and the presentation of products as they move along the supply chain. Packaging is distinguished in primary, secondary and tertiary packaging. Primary packaging comes in immediate contact with the products (e.g. a bottle), while secondary packaging can include one or more primary packages (e.g. carton, box) and it is used mainly for the standardization and the more efficient movement of the products across the supply chain. Tertiary packaging (or transportation packaging), refers to all those means (pallets, containers, etc.) that are used for the facilitation of the movement of goods and their protection against damages during the transportation process.

Packaging serves a wide range of targets, ranging from the protection of the product and the facilitation of its transportation to its promotion to the market. Consequently, the main axes of packaging are marketing, with the main goal of shaping consumer purchasing behavior and logistics, protecting the product during storage and transport by standardizing the corresponding systems.[34] Nevertheless, packaging constitutes an additional agent providing information about the product included, markings for special handling, storing conditions, ingredients, expiration dates and other important elements, operating as an advisor for the consumer, serving consumer needs from the food technology side.

In recent years, the increased interest on environmental issues, along with environmental risks deriving from the production, distribution and consumption of products, have led to the development of new approaches in labeling that allow the consumer to distinguish environmentally friendly from conventional products. A basic element of this new environmental policy that characterizes the entire global market is environmental labeling. In recent years, environmental labels (eco-labels) or ecological labels or environmental marking are a contemporary environmental policy tool that focuses on the role of information on the impact of production, distribution, consumption and disposal of products in the environment.[35]

The purpose of this chapter is to analyze this axis, namely environmental packaging, and in particular, environmental labeling.

Environmental labeling and labels

With environmental (or ecological) labeling, production enterprises express their environmental sensitivity and attempt to inform the consumers of the environmental consequences of their products. The ultimate goal is the adoption of a more environmentally friendly behavior on one hand, and the improvement of environmental specifications of products and services on the other.[36]

An environmental label is defined as a visual communication tool that presents products, services and businesses that operate based on specific environmental standards, distinguishing "green" from conventional products, thus influencing in a determinative way consumer as well as social behavior.

An ecological label is basically a label that determines the total environmental behavior of a product or service, based on its entire life cycle. The environmental behavior is guaranteed, based on the fact that the ecological label is provided by a neutral third body. Additionally, ecological labels are credible labels that are based on the life cycle assessment (LCA) tool, the methodology used to classify products, services and production systems based on their environmental characteristics. This means that they are attributed to an unbiased third party, essentially an environmental quality certification body operating on a non-discriminatory and transparent basis.

Thus, in the last 30 years, a significant number of environmental labels have been developed by private companies, organizations, non-governmental organizations (NGOs), etc. In particular, today it is estimated that there are more than 450 environmental labels used to mark products of almost any category, and their number is growing constantly.[37] For example, in the

European Union, environmental labeling is based on Regulation 880/1992 that focuses on promoting sustainable production and consumption. As a result, the community eco-label award scheme is awarded to stakeholders following a specific procedure. Consumers recognize the specific logo and identify products with the smaller environmental impact.

> Two very common examples of ecological labeling are the Energy Label and the Ecolabel Flower:[38]
>
> - The Energy Label communicates to the consumer the energy efficiency of home appliances. It also provides with information about energy or other resources consumption such as water.
> - The European Union's Ecolabel Flower communicates the most important environmental impacts on climate change, nature and biodiversity, energy and resource consumption as well as waste production and emissions of pollutants/harmful substances.

There are some essential features that a reliable eco-labeling system should have:

- Participation of an enterprise or organization in the scheme is voluntary; thus, it is not imposed by external laws or bodies. Also, these labels do not substitute the existing legislative framework. They just prove high environmental performance of an enterprise.
- Labeling must demonstrate that the product carrying it on its packaging has high environmental performance, higher than the average of products in the same category.
- Information entailed in environmental labeling is measurable and current scientific data. This data is provided to the user in a simple way and is updated regularly.
- Information entailed in environmental labeling is based on the assessment of the product's entire life cycle. All the phases of a product's "life" are being considered, from its design, production, operation and maintenance until its disposal.
- Issuing and assessing ecological labeling in general must be undertaken by a third-party, independent and reliable entity. Thus, the credibility, transparency and standardization of the evaluation procedure, certification and update will be achieved.
- The criteria, requirements and specifications of eco-labeling, as well as all parameters of the certification process, should be determined by committees whose members should represent the market – consumers, manufacturers, government bodies and consumer and environmental organizations.

Benefits and disadvantages of environmental labels

Numerous research initiatives have identified the benefits of the implementation of environmental labels (Figure 7.1):

- They promote sustainable production and consumption and contribute to environmental protection, achieving recycling and raw material savings, reduction in toxic substances use, replacement of dangerous materials with materials friendly to the environment and reinforcement of mechanisms supporting sustainable development in general.
- They are used by companies, as well as national and international organizations, to raise consumer awareness on environmental issues by recognizing and interpreting the impact of their actions and decisions on the environment.
- They mitigate the information gap between businesses and consumers regarding the environmental aspect of products. Specifically, they provide quantitative environmental information on the product that derives from credible methodologies like LCA.
- They support the consumer decision-making process. A product with environmental statement is highly competitive and marketable. Surveys have shown that "green" consumer behavior is more affected by the existence of environmental labels than the buyer's perceptions of the company and the strength of the brand. Therefore, the product is diversified against its competitors. In addition, consumers can compare

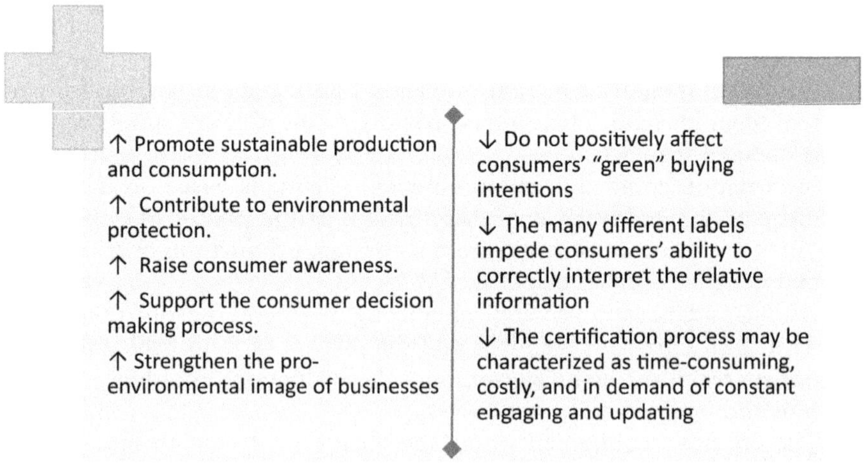

Figure 7.1 Benefits and disadvantages of environmental labels

products and choose the one with the smallest impact on the environment (performance) contributing to its protection.
- The (pro-environmental) image of businesses is strengthened, indicating environmental awareness and consciousness. Also, a reduction in economic costs and consumed energy is ensured by reducing the inputs used and the waste generated. In addition, the demand for environmentally friendly products by consumers is increased, and therefore, revenue is increased.

Disadvantages and challenges can also be distinguished:

- It is stated that environmental labels do not positively affect consumers' "green" buying intentions.
- Today, the existence of many different labels impedes the consumers' ability to correctly interpret the relative information. Usually, they have limited knowledge about labels and the different types of environmental labeling.
- The certification process may be standardized, though it often can be characterized as time-consuming, costly, and demanding constant engaging and updating.

The modern environmental labeling industry is characterized by the existence of a significant number of environmental labels that differ in terms of criteria and individual features.

Eco-label types

According to the International Organization for Standardization (ISO), environmental labels are categorized in three basic types: Type I, Type II and Type III, which are characterized by different principles and standardization processes.[39]

Type I environmental labeling – eco-labels (ISO 14024)

Type I environmental labeling, or simply Type I labels, refer to a product's environmental quality comparative to other products, and the goal is to raise consumer environmental awareness. These environmental labels are based on criteria that are set by third-party entities (standardization organizations, independent accredited bodies), that certify the use of labels for specific products and services that meet the relevant criteria based on the LCA method, such as the energy and water consumption, CO_2 emissions and waste management. In general, they aid in identifying those products that have lower environmental impact during their life cycle.

The standards and the criteria are usually defined in collaboration with various relative organizations and the certification is carried out through an independent process. The selection of products or categories of products, the definition of environmental criteria and the functional attributes of products as well as the assessment and certification procedure for acquiring the eco-labels are based on the international standard ISO 14024. Products meeting these specific criteria can use the eco-label for a definite period, provided that monetary obligations deriving from acquiring (fees and application costs) have been dealt with. Benefits of using these labels are obvious, but some weakness can be observed; for example, the fact that defining the criteria for a product category can take years. Moreover, for a large number of products, they are of limited use and their acquisition requires time and high costs.

Some examples of these labels are the EU eco-label, Nordic Swan, Blue Angel and the Energy Star, which are analyzed below.

- The EU Flower was established in 1992 and is recognized throughout the European Union and in other countries such as Norway, Lichtenstein, Iceland and others. The European eco-label is part of the EU's strategic goals to promote sustainable development. Since 2000, besides products, services are also included. The period of validity is limited until the renewal of selection criteria, while every member state of the EU is obligated to assemble independent bodies that grant the use of the European eco-label. Products that are eligible to acquire the eco-label fall into the following categories: cleaning products, clothing and textiles, bed mattresses and covers, paints and varnishes, floor coverings, electronic equipment, household appliances, furniture, lubricants, paper products and services like tourist accommodation.
- The Nordic Swan eco-label was established in 1989 by the Norwegian government. Besides Norway, it is recognized in Sweden, Finland, Iceland and Denmark. If a product is awarded with the label in one of these countries, the validity extends to the rest (given that manufacturers pay for the rights to use the label in the rest of the countries). The typical validity period is three years and on expiration, the manufacturers need to apply for renewal. If the criteria are stricter, or the product does not comply to them anymore, the right to use the label is lost.
- The Blue Angel was created in Germany in 1978 and it is the first eco-label introduced in the market. The label is property of the German Ministry of Environment and its management and promotion is carried out by the German environmental service and the national certification body. This is another case in which an independent body sets the environmental criteria and awards the label to those that meet them with transparent procedures.

- The Energy Star is a seal of quality carried by electronic office devices that meet low power consumption criteria. It has been established by the European Union within the framework of an agreement with the US government, with the goal of encouraging consumers to buy such devices to save on energy costs and protect the environment.

Type II environmental labeling – environmental claims (ISO 14021)

Type II environmental labeling, or Type II labels, are used to provide environmental information to consumers and other stakeholders and are not awarded or certified by an independent body. They are usually developed internally by the businesses themselves and take the form of a marketing statement (usually in the form of a logo, e.g. "100% recyclable product"). Hence, this labeling refers to a business' self-declared claim about the environmental performance of its products and is based on the ISO 14021 standard. Naturally, these claims must be accurate, understandable, clear, specific and relevant to the product and its use, and not untruthful or misleading. The manufacturers must be able to provide to any stakeholder the relevant information. In general, the lack of an independent certification body limits their usefulness and leads consumers to question the credibility of eco-labels altogether.

Type III environmental labeling – environmental product declarations (ISO 14021)

Labels of Type III are also named environmental product declarations (EPDs). They provide comprehensive environmental information, and are usually developed for specific products or product categories. They are based on bulk environmental data, such as the quantity of CO_2 emissions.

According to ISO classification, EPDs are voluntary certification programs that provide quantified environmental information of a product, based on predefined categories of individual parameters, defined and certified by an independent body. In particular, international standard ISO 14025 sets the main parameters of a product's Type III environmental labeling, which is based on independent predefined criteria. Additionally, possible environmental impacts are quantified. Impacts are assessed during the complete life cycle of a product with the LCA method, while the resulting data are validated by a third-party, independent body. Labels of this type are of high importance for trading between businesses (B2B), but they are of limited use in the consumers' market (for example, in the retail sector).

To make, file and publish an environmental statement according to ISO 14025, these steps are followed (Figure 7.2):

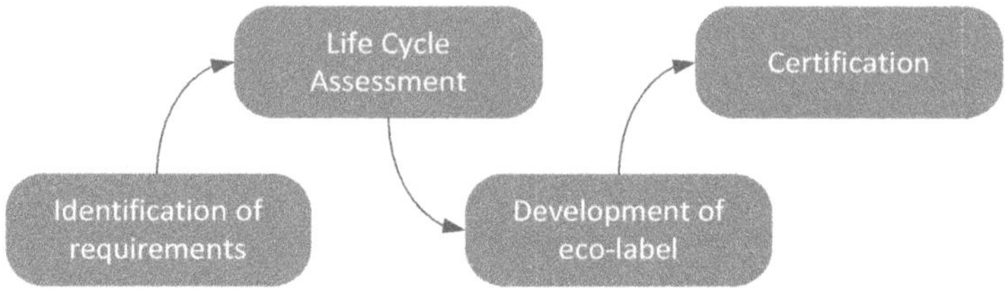

Figure 7.2 Environmental labeling process

Initially, a product category including competing products is selected and groundwork is laid for the specific product requirements. Data are then collected, and the calculations required for LCA are carried out to determine the prominent environmental impacts from the product's use. Strict environmental criteria for the product category are set. Information on applying the environmental label is collected, and finally, the company is awarded with the environmental label. Only products or production methods that comply with these criteria are eligible for being awarded the eco-label.

Committees made up of government officials, businesses, academic institutions, consumer associations and non-governmental organizations choose product categories and set environmental criteria, while ensuring transparency and representation of all stakeholders. Both the criteria used and the eco-labels awarded to the products have a limited duration. Over time and as the number of products awarded the eco-labels grows, the process becomes more and more stringent in order to preserve the credibility and validity of the eco-label. Typical validity duration of an eco-label is two to five years. After this time, the scheme of the label is enriched with the new scientific and technological facts in order to enhance the environmental performance of products bearing the labels.

In addition to this standardization, other types of eco-labels that are essentially hybrid environmental labels exist and share common features between the different sub-types. Therefore, they cannot easily be assigned to a particular category of the ISO system.

Environmental labeling applications

Environmental labels are used in the following functional areas (Figure 7.3):

- *Green procurement:* Both at management and product levels, environmental information from green procurement is a proven useful tool. Information related to management can be obtained from data coming from environmental management systems of suppliers and environmental reports – or simply, an eco-label is used to document the environmental requirements of suppliers.
- *(Eco) Design of new products:* Data collected to obtain environmental labeling can be utilized to design new product processes or to modify old ones to reduce their environmental impact. In the case of designing new products or services, eco-labels are a very useful tool for eco-design by selecting all those elements and materials that contribute to the least possible environmental impact.
- *Supply chain relationships*: Information derived from environmental labels is objective, reliable and based on real data. This tool makes transactions between businesses easier; cooperation between them is facilitated, for example through green procurement.

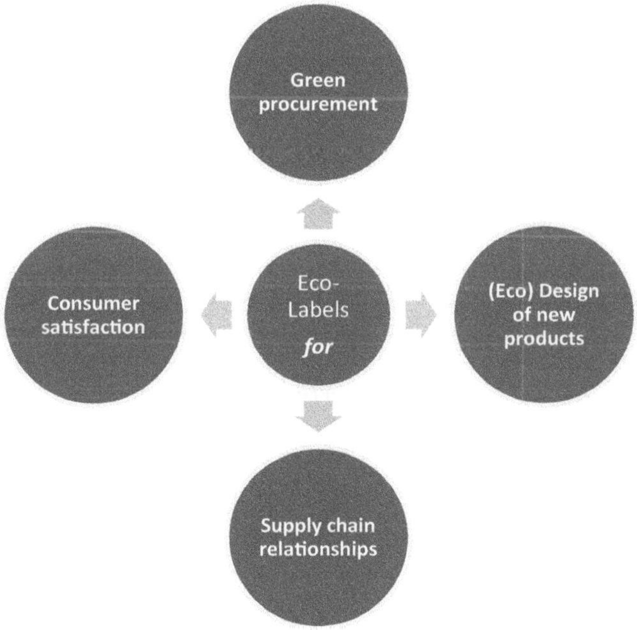

Figure 7.3 Environmental labeling applications

- ***Consumer satisfaction:*** Environmental labels are primarily designed to encourage consumers to an eco-friendlier way of life. They achieve this by providing the appropriate information. For example, in agri-food products, the amount of CO_2 released into the atmosphere through the activities of the supply chain member companies is displayed.

> The agri-food sector includes both primary production from agriculture as well as the wider food industry using primary products for food production (processing, marketing, transport, distribution). The need to address environmental challenges such as the high rate of biodiversity loss, climate change, etc., has led to the growing demand for high-quality agricultural food products. The environmental impact of food production is a complex system and can be traced throughout the production, processing, processing, distribution and end-use chain.
>
> In Greece and in the European Union, various tools and structures have been developed, with the help of which quality can be certified, i.e. to document compliance with specific requirements and characteristics of either the product itself or its production process. Towards this direction lies the acquisition of an environmental label, which is a voluntary method of environmental certification to promote sustainable production and consumption.

Chapter summary

In this chapter, we focused on green packaging and the role that it plays in everyday life. Environmental labeling was analytically discussed, and labels that exist in market were highlighted. The essential features that any reliable eco-labeling system should have were presented and the benefits of environmental labels were emphasized, together with their disadvantages. The types of environmental labels were discussed, and the main functional areas where those are mostly used were identified.

8 Reverse logistics

Discussion questions

- What are the available alternatives for the management of solid wastes?
- What is integrated solid waste management, and how is this realized in the case of waste electrical and electronic equipment or construction and demolition waste?
- What are the key considerations of waste management within the sustainable development concept?

Introduction: waste management and environmental policy

One of the main factors that pose significant pressure on the environment today is solid waste. The volume of the waste generated is constantly growing due to the increase of the development rate, but also due to change of consumer patterns. After cement production, urban solid waste represents the largest mass of solids produced worldwide. At the same time, however, public awareness is growing, as citizens' demand for sound environmental waste management is becoming more and more intense. For the integrated management of both the total amount of waste, as well as each individual stream, the organizational scheme is required to be designed in such a way to ensure – in order of priority – the following (Figure 8.1):

- Minimization of waste production, by encouraging the reduction of their production.
- Reuse of materials.
- Recycling of materials.
- Recovery of energy, in special facilities, with the production of electrical and thermal energy.

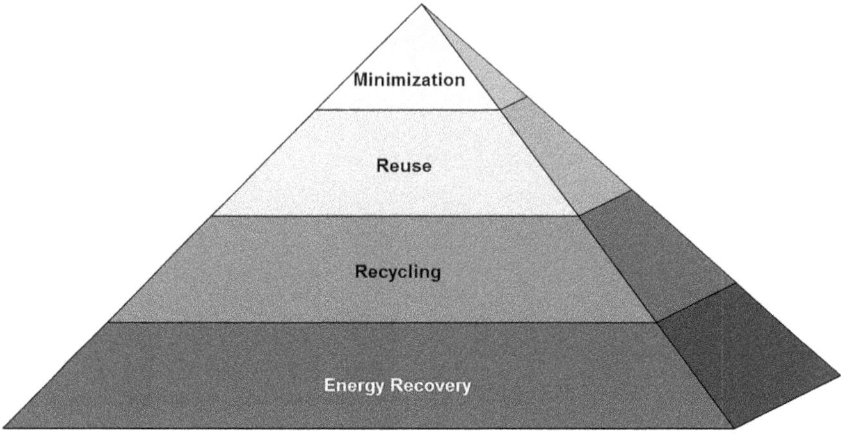

Figure 8.1 Hierarchy of waste management alternative strategies

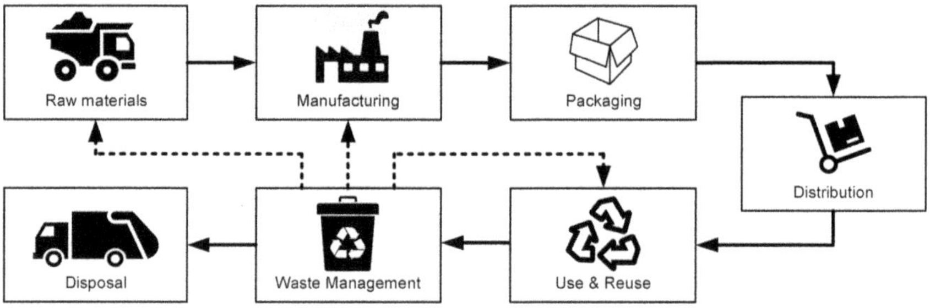

Figure 8.2 Product's life cycle

The European Union's environmental policy on the management of municipal solid waste is based on these four pillars. Prevention, which is considerably preferable to corrective actions, is of primary importance. A key issue in the prevention of waste production is the assessment of impacts throughout the life of a product, from the stage of raw material extraction to its final disposal (Figure 8.2). In order to prevent waste production, incentives and/or disincentives, as well as the promotion of social sensitivity and encouragement of consumers to buy less-polluting products (e.g. through the promotion of the eco-label), are particularly important. At the same time, prevention can be achieved by imposing restrictions or even prohibitions on the use of certain substances to prevent the production of hazardous waste at a later stage.

In addition, European environmental policy aims at recycling and utilization of materials.[40] The recovery of materials from waste is the overarching goal of any sustainable policy making. In cases wherein waste cannot be avoided, disposed products should be reused or subjected to material recovery procedures. A basic parameter for the recovery of materials is their separation at the source, a practice that involves the active involvement of consumers and end users in the management chain. In cases where materials cannot be recovered, heat treatment is proposed for energy recovery.

At the same time, the "polluter" is considered responsible for the cost of the interventions required to protect the environment. Based on the extensive responsibility of the producer, the manufacturer of a product becomes responsible for securing the resources and means to ultimately manage the waste products thus placed on the market. This poses incentives for the producers not only to minimize the waste generated by their products, but also to optimize their design in a way that facilitates their re-use and/or recovery at the end of their useful life.

In addition, it is worth emphasizing the principle of energy recovery from waste prior to the corresponding alternative of safe disposal.[41] Due to the relative economy of the disposal of waste in landfills, this alternative of final disposal is the most common practice today at the international level. However, environmentally it is considered to be considerably inferior to management modes such as reuse, recycling and heat treatment, and should therefore only be chosen as a last resort in cases when alternative management is not feasible. It is worth noting that despite the scientific community's contradictory estimates that heat treatment and recycling are conflicting practices, in most countries heat treatment of urban solid waste is in line with recycling. In contrast, in countries where landfill is the main practice, recycling rates are also low.

Due to their significant volumes, the cases of reverse logistics for waste electrical and electronic equipment (WEEE) and construction and demolition waste (CDW) are further detailed and analyzed in the following chapters.

Case study: waste electrical and electronic equipment (WEEE)

The waste streams

Electrical and electronic equipment (EEE) is defined as "equipment whose correct operation depends on an electric current or an electromagnetic field and is designed to operate at rated voltages of up to 1,000 V AC and 1,500 V DC".[42] In general, the production of EEE is considered one of the fastest-growing businesses in the world. Modern technological innovations and the expansion of the EEE market speed up the replacement of EEE representing dated technology, thus leading to a significant increase in waste EEE (WEEE), with annual growth rates ranging between 3% and 5%. As a waste stream, WEEE are heterogeneous and complex in terms of materials and ingredients. Their complexity is due to the wide variety of materials used as raw materials for the production of EEE products, and a feature of this particular waste stream is that many of these materials are considered to be particularly toxic.

The European Union has introduced extensive environmental legislation for WEEEs, which is based on the responsibility of producers who are required to finance their collection, treatment, recovery and environmentally sound disposal. In summary, although there is still a significant lag, in recent years remarkable progress has been made towards the sustainable management of the special streams of WEEE in Europe.

Concerning the quantities of WEEE produced today at the international level, a significant number of studies have been carried out that present conflicting results, since they differ significantly due to the different methods and assumptions that are taken in each of them. Generally, it is estimated that according to the modern consumer habits, each person produces about 12–20 kilograms per year of WEEE.[43] Taking this assumption into consideration, the volume of the total yearly quantity of WEEE is considerably high, which in combination with its hazards makes it one of the most important waste streams in the world. At the same time, as has already been mentioned, their quantity is constantly growing at a particularly intense rate, mainly due to the rapid development of technology, the use of EEE products in more and more areas of human activity, but also in the more frequent use of automated procedures in production processes. Their growth rate is up to three times higher than the corresponding growth rate of municipal solid waste.

Despite the increase in the quantities of WEEE which has been recorded internationally over the last decade, the quantities still correspond to only a small fraction of the total quantities of the WEEE produced. Notwithstanding the legislative regulations for the obligation for separate collection and processing of WEEE, unused EEE products – mostly small appliances – continue to be discarded together with other municipal solid waste. In a relevant study, it was calculated that WEEE accounts for about 1% of urban waste.[44]

Figure 8.3 presents the typical composition of the WEEE stream.[44] Of the total WEEE quantity, about half are ferrous materials, which demonstrates the recyclability, but also the remaining WEEE value at the end of their life cycle. Ferrous metals are highly concentrated in IT and telecommunication products, but also in small home appliances. Despite ferrous materials, the disposable EEE products have significant value – in terms of recycled materials – because of their composition in copper, aluminum and glass.

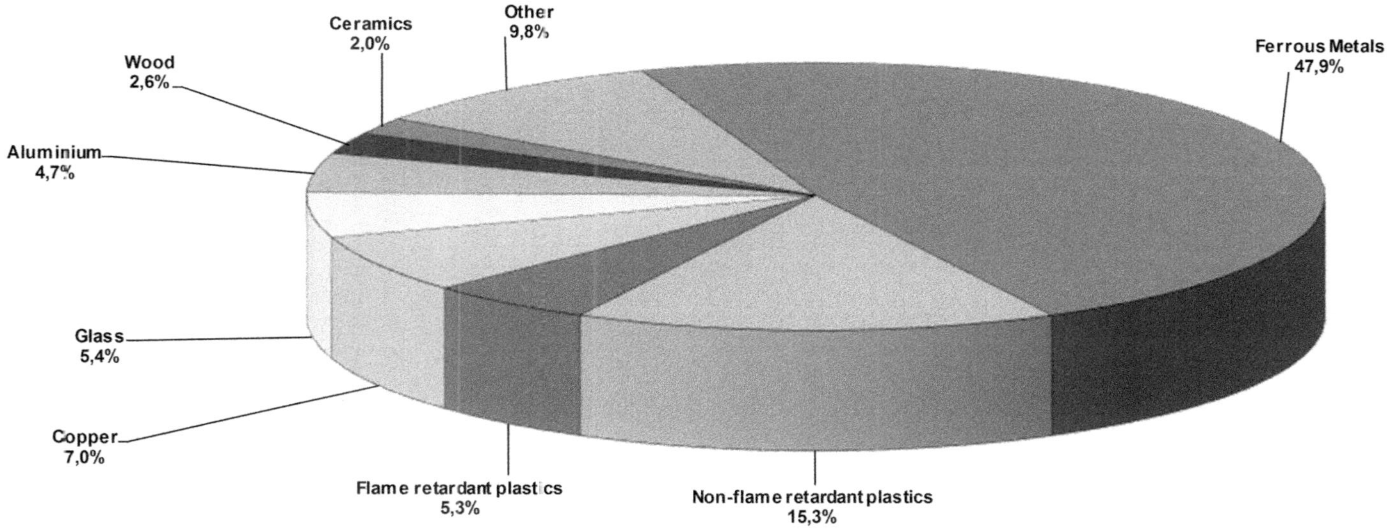

Figure 8.3 Typical composition of WEEE

Non-ferrous metals are mainly found in "brown products", control devices and medical equipment. Glass is located in the WEEE, mainly in lighting fixtures. It is worth noting that while recycling of glass is relatively easy, the high percentage of contaminants (metal oxides) found in WEEE makes it particularly difficult to process. The second largest fraction of materials in WEEE, after the ferrous ones, are plastics, which represent more than 20% of the total quantity. The largest volume of plastic from WEEE comes from large household appliances, while high concentrations are found in both small household appliances and telecommunications equipment. Of these, non-flame-retardant plastics account for 15% of the total, while those with flame retardants, which require special management, account for 5% of the total WEEE stream.

Apart from the materials already mentioned in the WEEE stream, there are also precious materials, which, although in very small quantities, are an important component of the residual economic value of disposable EEE. An example is the case of gold in mobile telephony products, that although its share is far less than 1%, the total economic value of its recovery exceeds 65%.[45]

In addition, EEE products include materials that despite their small quantities are especially dangerous for the environment and public health. Such materials are cadmium, lead, mercury, hexavalent chromium, nickel, brominated flame retardants, P.V.C., polychlorinated biphenyls (P.C.B.s.), heavy metals, liquid crystals and others. Compared to the usual urban solid waste, EEE products contain significantly higher quantities of bromine and lead, as well as heavy metals.

The environmental hazard of these materials depends on their toxicity, as well as on the quantities released during WEEE management. Practically, the most important environmental risks come from materials and components, such as (a) electronic circuits and mercury from the old equipment; (b) halogenated flame retardants in plastic cases, cables and printed circuit board systems; c) liquid crystal displays; (d) cadmium in nickel-cadmium batteries; and (e) lead oxide in cathode ray tubes. Also, memory-based processors identify components containing substances such as beryllium oxide and arsenic. Therefore, during disassembly of WEEE, hazardous materials are produced which require special treatment methods for their proper environmental disposal. Table 8.1 summarizes the main materials and the proposed methods of disposal.

In addition to the undoubtedly increasing volumes of WEEE, at international level, their composition is expected to change significantly over the coming years. This is expected to happen both because of the technological developments in the field and due to the relevant legislation, according to which the use of certain hazardous substances in the EEE products is prohibited (Restriction of Hazardous Substances [RoHS]), and because restrictions are placed on the eco-design of energy-using products. In the future, due to the already existing and imminent changes in the design of

Table 8.1 Hazardous materials in EEE products and disposal methods

Materials	*Process method*
Capacitors containing PCBs	Heat treatment
Electrolytic capacitors containing dangerous substances	Utilization of metals, heat treatment
Batteries (unsorted)	Final disposal as hazardous waste
Lead battery cells	Utilization
Nickel-cadmium battery cells	Utilization
"Button" batteries	Utilization
Printed circuit boards	Utilization, heat treatment
Cathode ray tube coatings	Final disposal as hazardous waste
Liquid crystal monitors	Final disposal as hazardous waste
Components containing mercury	Utilization
Components containing asbestos	Final disposal as hazardous waste
Containers of liquid or sticky inks as well as colored	Reuse, utilization, heat treatment
Gas discharge lamps	Final disposal as hazardous waste

EEE products. entering the market, WEEE is expected to include fewer quantities of ferrous and precious materials, while the amounts of aluminum and plastic are projected to increase significantly compared to their current levels.

Available assessment alternatives

From the typical composition of the WEEE, it can be understood that this specific waste stream is a particularly complex mixture of materials and components that on one side have significant residual value, but on the other present a significant environmental risk. To that end, its final disposal and management requires special care from the organization undertaking it. For the EEE assessment needs at the end of their life cycle, the coexistence of various final disposal methods is necessary along with the adoption of an integrated strategy, which must include the re-use of products where possible, recycling for the recovery of materials, heat treatment for energy recovery and even landfill for the final – safe – disposal of the residues of the above processes. Figure 8.4 illustrates the stream of a typical EEE product at the end of its life cycle. In practice, apart from the processes depicted in Figure 8.4, there is still the export – both legal and illegal – and deposition of the WEEE in Third World countries.[46]

Especially regarding the recovery of materials and energy from the WEEE, Figure 8.5 presents the sequence of operations that is followed. The first step is the separation of ferrous and non-ferrous metals – such as copper and aluminum, which are included in the WEEE in significant quantities – and then both streams are led to recycling. The residues resulting from this separation, along with the significant quantities of plastic that cannot be recycled,

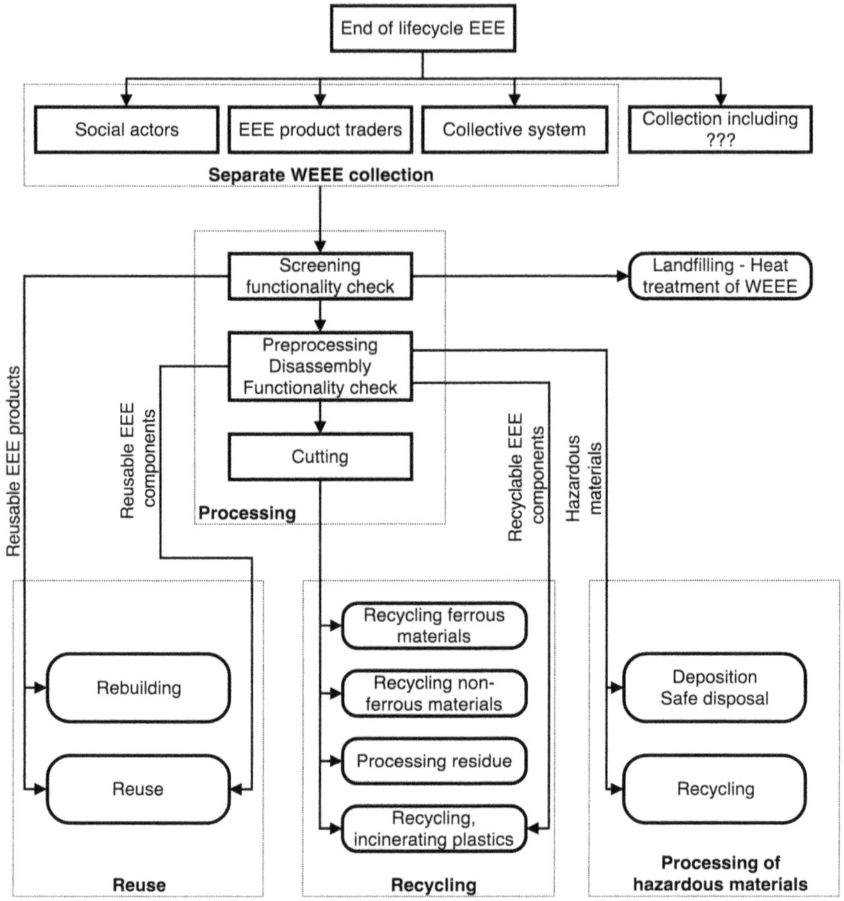

Figure 8.4 Stream of EEE product at the end of its life cycle

can be taken to either a landfill or a modern Thermal Treatment Unit (TTU) in order to recover significant amounts of energy.

The following sections summarize the main features of the individual alternative management modes that make up an integrated management system of the WEEE. It is stressed that none of the methods can, by itself, be a solution to the management of the WEEEs, since they all have advantages and disadvantages over the others. In addition, there are limitations regarding the WEEEs, or the parts of them that can be managed. It is important that the individual alternatives form a part of a framework for the integrated management of the special WEEE streams.

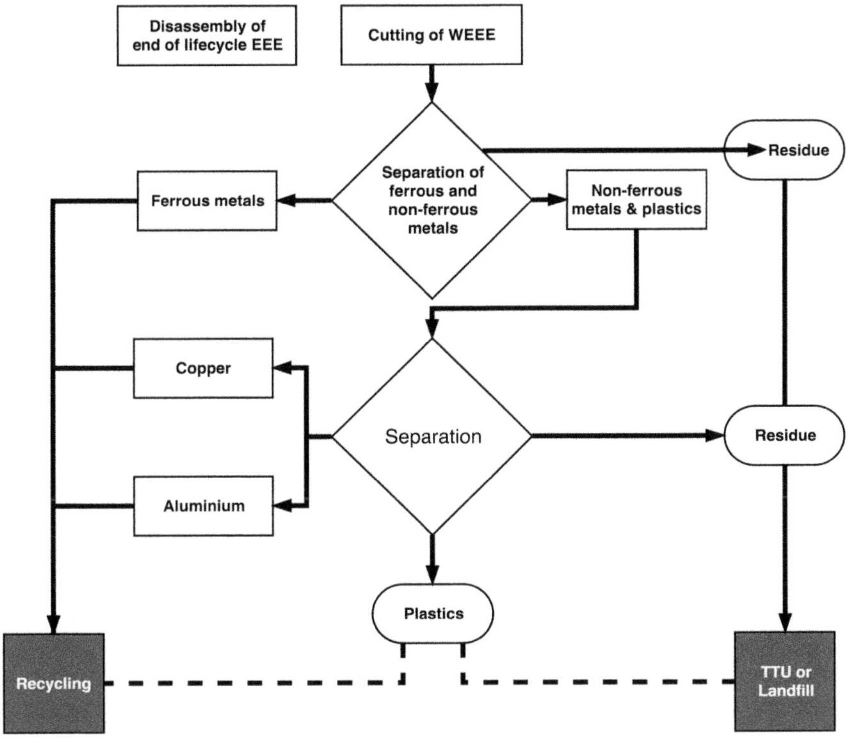

Figure 8.5 Stages of material recovery from WEEE

Reuse

Reuse is the process during which a product is used from the beginning in the same or similar purpose for which it was originally manufactured, using its original shape. The reuse of EEE products when they have reached the end of their life cycle is generally considered as first priority in an integrated WEEE management system. The main objective of this practice is to maximize the product's lifespan as much as possible, thereby reducing its average burden on the environment. Reuse can be accomplished either by reusing the product as it stands or by repairing and reforming it. Also, if it is not possible to reuse a product as such, its dismantling and re-use of individual parts or even its components should be considered.

Despite the preference of the reuse method as a hierarchical best practice for the management of WEEE, there are some issues that need to be considered. First of all, an objective problem for the promotion of this alternative is the relatively high cost of the manual dismantling of a product – which is the basis of the particular WEEE management process – when most of the components have a relatively low price on the international electronics market. Also, the wide variety of EEE products does not allow for the adoption of common management practices, thereby significantly increasing the cost of re-use.

Recycling

Material recovery through recycling is one of the most fundamental alternative choices for WEEE at the end of their life cycle, considering that reuse – which is hieratically preferred to recycle- is especially difficult because of the rapid technological evolution and the following devaluation. The large amounts of WEEE contain among others significant quantities of plastic, glass, ceramic and metal. The process methods used for recycling WEEE are classified in three basic categories:

Cutting/Separation: This practice is the most prevalent recycling method at the international level. The basic principle of this practice is the fragmentation of EEE products in very small pieces, aiming to reduce their volume and thereby improve the homogeneity of the resulting pieces. Thus, the separation of the various materials by suitable processes is facilitated in order to recover the materials. Separation can also be described as uncontrolled disassembly. In summary, the unused EEE products are introduced into a cylindrical container – in some cases first, they are compressed in presses – where they are cut by rotating hammers until they reach an appropriate size, usually ranging from 10 to 100mm. The processing sequence of the cutting – separation practice is given schematically in Figure 8.6.

More specifically, after the WEEE cutting their separation follows, which takes place in three distinctive stages: electrostatic separation, air separation and inductive separation. In electrostatic separation, the magnetic capability and the electrical conductivity of the materials are used for the separation.

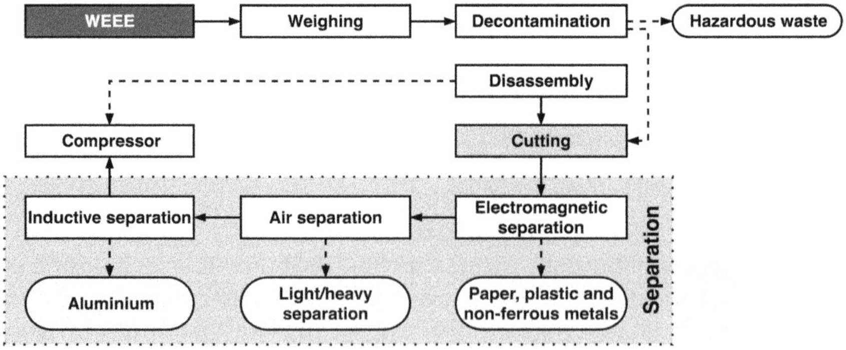

Figure 8.6 Stages for WEEE material recovery process

Typically, electrostatic separation is used to separate paper from plastic and non-ferrous metals from other waste, such as separating copper from plastic cables. Air separation is where a mixture is classified into various streams of materials under the influence of air. The separation is based on the different particle trajectories within the air layer and gravity. Separation is achieved depending on the weight, humidity and shape of the particles, as well as the velocity of the blowing air and the time spent in the air separator. As a form of air separation, specific weight separation can be considered, which takes advantage of the differences in the specific weight of the different materials in the common fraction, using a cyclone or centrifugal separator. Specific weight separation is applied to the separation of heavy-light fractions, such as steel and plastic. Inductive separation is where the separation and recovery of aluminum is achieved by using eddy currents. The process is completed by compressing the waste stream using a compressor. With the compressing, chipping and briquetting machines, surface reduction, density increase, and easier handling of materials are achieved, thus creating the ideal storage and transport conditions.

Metallurgical treatment: This practice mainly uses non-metallic components of the products as fuel for melting and recovering metals. In the metallurgical treatment, the precious materials are disassembled and the liquids are isolated, while the rest of the stream is compressed and led to a blast furnace. There, the metals are melted and separated appropriately, so as not to create undesirable alloys. The type of this treatment depends significantly on the requirements of the recycled materials, and more specifically on their purity levels. Non-ferrous metals, such as copper, can form alloys with molten steel and thus significantly and permanently reduce its value, since copper is very difficult to remove from steel alloys. Regarding the value of this particular recycling practice of WEEE, many are the supporters of the method as they consider that the overall energy benefit deriving from the whole process is far more preferable than product dismantling and utilization of their materials. Nevertheless, there is a significant portion of critics of the method, who simply regard it as a covered method of incineration of WEEEs.

Disassembly: The method of disassembling the components of an EEE product is distinguished between non-destructive and partially destructive. During the first, the individual components of the EEE waste products are not destroyed, while in the second the elements are destroyed, with the most common destruction of their connections. Today, the disassembly process has not become fully automated, other than specific minimal experimental exceptions. At a practical level, automated disassembly of EEE products has not so far had the expected development for a number of reasons, such as the increased investment costs required, the particularly high product variety and materials contained in the WEEE, and the lack of construction data for the WEEE. Nevertheless, it is certain that in the future such processes will be automated as a result of the environmental design of EEE products, which will significantly increase the viability of this alternative. Manual disassembly and separation of WEEEs in its components significantly increases processing costs, so it is performed only for the valuable materials contained in the unusable EEE products.

During the recycling process, the toxic or dangerous substances contained in the EEE products can be released, with adverse effects on the environment, but also on the safety and hygiene of people working in recycling plants. For example, when processing and recycling cathode ray tubes or electronic circuit boards, toxic metal and organic substances are released as fine molecular powders and smoke. Also, when heating plastics containing flame retardants – bromine or chlorine compounds – prior to their recycling, dioxins and furans are formed which are particularly harmful to public health. It is therefore necessary to take measures that will protect both human health and the natural environment. In many cases, proper handling and controlled pre-treatment of WEEEs can minimize the risks associated with these substances. Typical is the example of controlled crushing of fluorescence lamps for the retention of mercury emissions that prevent uncontrolled breakage and emissions during transportation to the WEEE management facilities. Similar preventive measures help to ensure that recycling can be a good alternative to the final disposal of EEE products.

Heat treatment

Hierarchically, energy recovery is the ultimate alternative for waste stream management. Incineration in special installations is mainly considered as recovery of energy where the use of the energy emitted by combustion is allowed. Residues resulting from the recycling process of WEEEs is possible to be exploited to recover energy. Thermal treatment reduces the toxic or dangerous activity of some substances while increasing danger of others, such as polyvinyl chloride (PVC).

Generally, emissions of toxic or dangerous gases are attributed in the heat treatment of WEEE, while the most important problem of the incineration of WEEEs is the emissions of dioxins and furans originating from the incineration of halogen-containing plastics, as bromine and chlorine are more common as flame retardants in the plastic parts of the EEE products.

From the total WEEE stream, the fraction with the greatest energy utilization potential, are the plastic parts, which in larger quantities are used as external cells or EEE products. Today, plastic is becoming an even more important part of EEE products, while predictions are that their percentage will continue to grow in the following years, especially in high technology sectors such as telecommunications. Given the significant calorific value of plastics, which is higher than that of carbon and at least twice that of wood and paper in urban waste, it becomes clear that the plastic parts resulting from the residuals of the WEEE recycling process are a great source of energy recovery.

Final disposal

The disposal of WEEE in landfills is the least preferred environmental choice regarding their management in the end of their life cycle. The fact that EEE products contain many harmful substances results in adverse effects on the environment during their internment. To limit such impacts, there should be controlled landfills meeting specific technical and environmental specifications. Given the difficulty of completely sealing a landfill throughout its life, there is a risk of leakage of heavy metals and other hazardous chemicals to the environment, particularly in the underground soil and waters. For example, mercury leaching is observed during the destruction of circuit breakers, while in the case of destruction of PCB-containing capacitors or during the internment of EEE containing plastics with brominated flame retardants or cadmium, these toxic substances can end up in the ground and underground waters. Environmental impacts are undoubtedly far more severe in cases where EE products are thrown uncontrollably into the environment or in places where technical and environmental safety standards are not met.

Integrated WEEE management

The development of a methodological framework for the integrated management of the WEEE is necessary to integrate all the components and stakeholders involved problem and to identify actions that need to be undertaken by both the manufacturers of the EEE products and the state. According to the "polluter pays principle", the manufacturer (or importer) of EEE products is responsible for the environmentally sound management of its products when they reach the end of their useful life. Simultaneously, the role of the state is important, and it should be able to provide all the necessary infrastructure that is necessary for the proper operation of an alternative management system for this particular waste stream. For this reason, in the methodological framework developed in this section, the analysis of the proposed actions takes place on two distinct, complementary levels:

- Actions required from the manufacturer's side for the improvement of the assessment of the unused EEE products he is responsible for.
- Actions required from the state's side to provide the necessary infrastructure for the development of the reverse supply chain of EEE waste products, as well as the optimal organization of the alternative management system of the WEEE.

Figure 8.7 illustrates the main elements that compose the methodological framework for integrated management of the WEEE stream. As part of the extensive producer responsibility, from the EEE manufacturer point of view, it is proposed to promote the alternative management of the products they

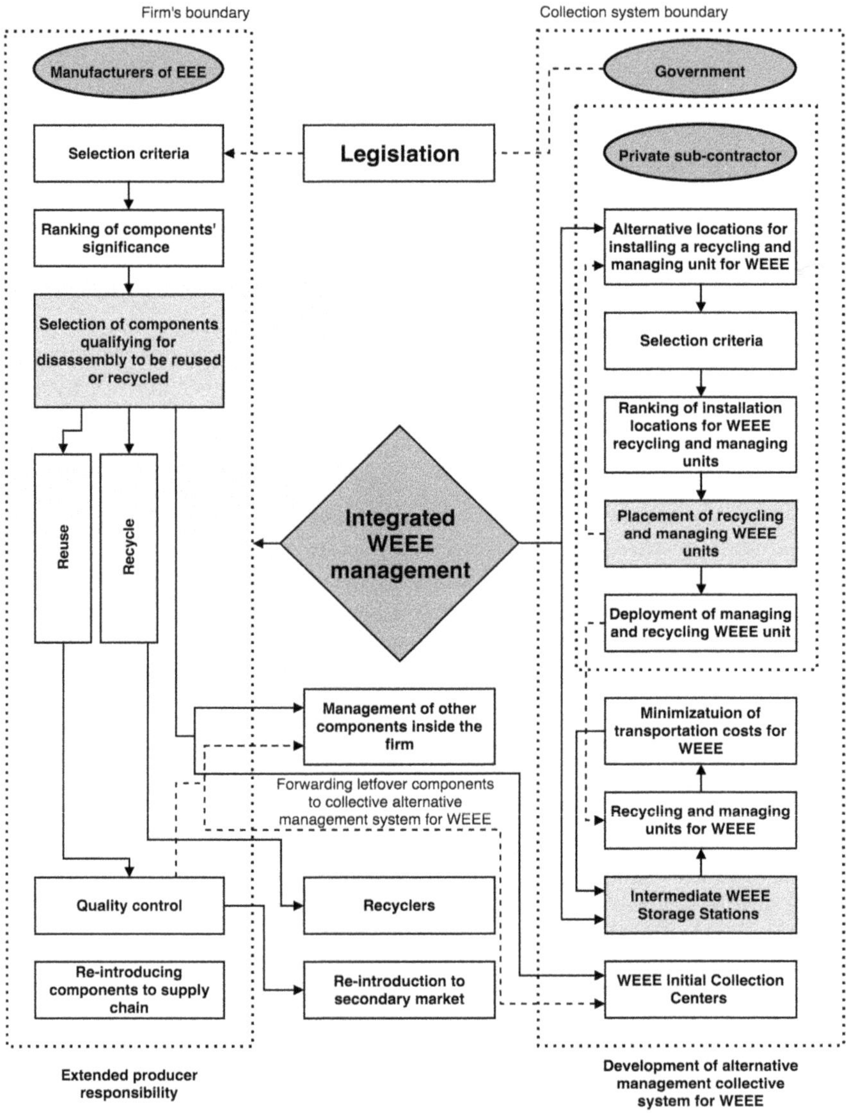

Figure 8.7 Methodological framework for the integrated assessment of WEEE

place on the market by reusing – either by reintroducing products and/or components in their supply chain, or by promoting them in the secondary market of the EEE – and recycling on the other. Other products that are not eligible for alternative management are promoted either to an alternative management system for WEEE or to a disposal infrastructure that the company itself can provide. The critical parameters that have to be taken into consideration in order to develop the collecting system concern the required processing and recycling facilities for the particular waste stream that meet national requirement as well as the minimization of the transport cost of the reverse supply chain of the EEE products by optimizing the location of collection points, interim storage and WEEE final disposal.

The most critical point in multidimensional management problems is the offsetting of different kinds of available information that can eventually lead to a solution. The development of a methodological framework for integrated management of WEEE is a particularly complex process if it is necessary to evaluate a range of alternatives, which may seem likewise equivalent or incomparable. While documenting, both in the practices of managing unnecessary EEE products and in the characteristics of an integrated alternative management system for WEEEs, a series of critical parameters is involved. The combination of all the parameters makes the development of a methodological framework, which includes and takes into account all the individual axes of the integrated management of WEEE, a complex problem. The final choice of the best solution between alternative scenarios requires the reconsideration and evaluation of many conflicting parameters and socio-political, economic and environmental nature objectives.

The proposed methodology adopts multi-criteria analysis techniques. In this way, a systematic and mathematical standard solution for problems arising from conflicting objectives is made, posing a conciliatory effort among them. Multi-criteria analysis is an important tool whose main objective is to assist the decision-maker to prioritize all available techniques, taking into account a critical number of criteria.

Multi-criteria WEEE assessment from the manufacturer

In line with standard industry practice for EEE products, for the development of an industrial product, particular attention is paid to construction costs, without taking into account environmental parameters and environmental planning issues. This is contrary to the requirements of sustainable development promoted internationally in recent years, whereby cost considerations should be taken into account, along with environmental and social considerations (Figure 8.8).

More specifically, the strategic objective of an EEE producer should be a mix of policies such as social corporate responsibility, eco-efficiency and environmental friendliness. This objective is also in line with the three main principles of the integrated product policy (IPP), a holistic approach to the

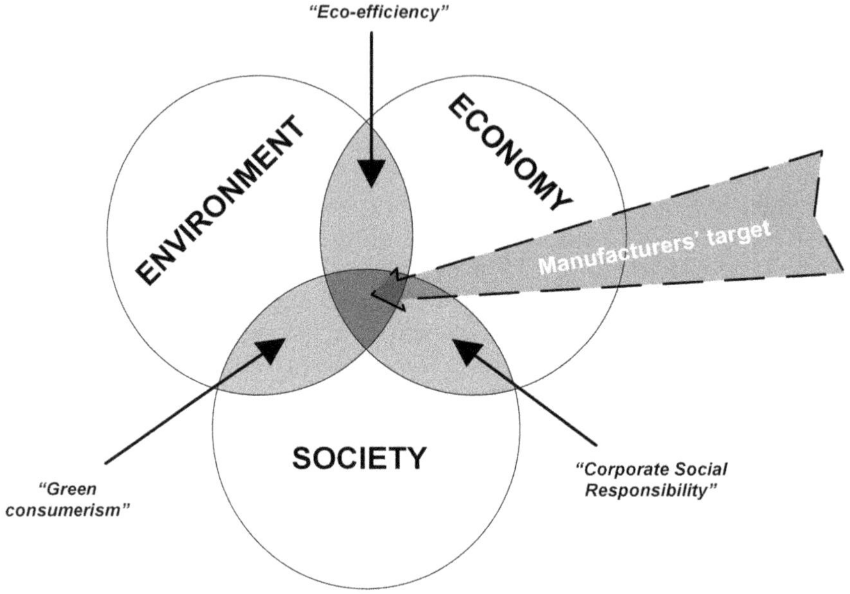

Figure 8.8 Key considerations within the sustainable development concept

problem, impact assessment throughout the product life cycle and the satisfaction of market needs. In addition to legislative and institutional interventions, for manufacturers of EEE products, there may be additional economic benefits from applying the principle of extended producer responsibility as demonstrated by a number of successful practices at the international level.

An equally important parameter that should be taken into account when adopting an integrated management policy for the WEEE from manufacturers of EEE products is the "turning" of the consumer community into more environmentally friendly products. At the international level, consumers begin to feel responsible for the environmental behavior of goods and their possible contribution to environmental protection, resulting in a preference for environmentally sensitive businesses investing in "green" policies and technologies. This growing trend towards "green consumerism", characterized by consumer preference for eco-products and their acceptance of paying an additional amount for a product or service that causes less environmental burden, is also a powerful incentive for businesses.

It is worth noting that in an industrial product, all its components do not have the same weight in terms of their economic and environmental burden. Pursuant to the Pareto law, also known as the "80/20 rule", in the case of an EEE product, a very small percentage (about 20%) of all components is responsible for most (about 80%) of the burden, both in economic and environmental terms. For this reason, it would be of great interest for manufacturers of industrial products to be able to identify components – elements within a product, which would bring maximum benefit if they were properly managed at the end of the product's useful life, either at an economic or environmental level.

The methodological framework outlined in Figure 8.7 takes into account environmental and economic criteria as well as environmental planning issues in choosing the optimal management methodology of an industrial product in the end of its life cycle. The main objective is to adopt alternative forms of management, more environmentally friendly, if not for the whole product, at least for those components which either have a large share of the overall environmental burden of a product or have significant residual value for the manufacturer. In particular, to determine the optimal way of handling the individual components of an EEE product at the end of its life cycle from the manufacturer's side, the following individual critical parameters should be taken into account: (a) component procurement costs; (b) component environmental burden; (c) component weight; (d) quantities of identical parts in the product; and (e) ease of disassembly of the component.

These parameters are combined with each other using weighting factors to calculate the final ranking of the "significance" of each component in relation to the remainder of the product concerned. The objective of such an approach is twofold. On the one hand, an attempt is made to prioritize the components of an EEE product in order to identify those with the highest value at the end of the product's life cycle since, as already mentioned, only a small percentage of the total is responsible for the overwhelming majority of environmental and economic impacts (Pareto law, Figure 8.9).

In this way, the effort to improve the overall behavior of a product is more targeted and therefore more effective. On the other hand, with this

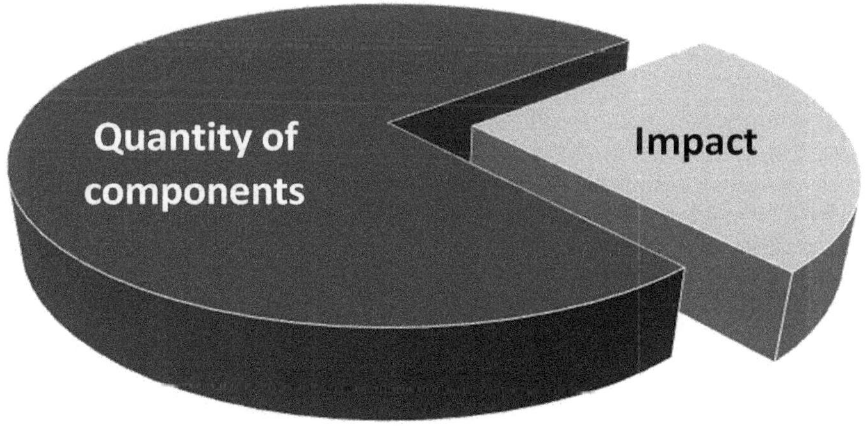

Figure 8.9 Application of Pareto law in WEEE management

approach, the optimal method of managing the individual components when the product in which they are contained reaches the end of the useful life is indicated to the manufacturer of EEE products. This approach identifies the components to be promoted for alternative management (reuse or recycling) and separates them from the non-interest components and can be sent for final disposal or incineration in appropriate units.

By adopting and applying this approach, it is possible to rank the components in terms of their "significance" at the end the useful life of the product under consideration, and on the other hand, determine the optimal method of alternative management of the individual components. Additionally, this methodology substantially contributes to the redesign of the EEE products under examination, as critical parts that require particular attention have already been identified. In this way, for those components that have either significant residual economic value or are responsible for a significant proportion of the overall environmental burden of the product under consideration, it is proposed to further investigate the redesign of the device, aiming at its subsequent optimal disassembly.

Based on the logic that disassembly is not reverse assembly, you should not only consider the design for easy assembly as is currently the case, but also the design for easy dismantling. Many times, the two processes have conflicting "requirements", so it's not always easy to reconcile them. The logic of redesign is based on the immediate disconnection of the components selected as the most relevant for the overall behavior of the product under consideration during the disassembly phase. The redesign proposed primarily concerns the type of connection of the components to each other in such a way as to facilitate, as much as possible, the dismantling process, considering the specific characteristics of the product. At the same time, redesign should take into account the tools to be used for dismantling in a way that promotes their greatest possible simplification by promoting newly integrated dismantling processes. It is also possible to propose a modification of the structure of the product to facilitate its disassembly.

In addition to minimizing the complexity of disassembling an EEE product, redesign may take another dimension, that of promoting the recycling of its components as opposed to burial or burning. The design for recycling complements the design for disassembly. Thus, manufacturers can intervene in the original design of the product either by using recyclable materials or by introducing recovered materials and components into their production process in order to strengthen the secondary market. Individual actions proposed to promote design for recycling are: (a) minimizing the variety of materials; (b) easy accessibility of hazardous and/or valuable components; (c) distinction between the same materials (e.g. the same color); (d) the avoidance of the use of hazardous substances or materials and elements incompatible with standard recycling processes; and (e) the reduction of recyclable plastic and composite materials.

Of course, the integrated management of the WEEE stream requires a harmonious cooperation and linking of the relevant actions of all the management actors involved. Consequently, action is required to provide all the necessary infrastructure for the processing and recycling of EEE products from the state's side. In this direction, it is necessary to develop and operate a collective alternative management system, responsible for the development of all the necessary infrastructure and the coordination of the collective actions by the state. These infrastructures refer – among other things – to the development of WEEE initial collection centers and Intermediate WEEE storage stations to achieve scale economies in order to minimize the transportation cost of the reverse supply chain, which is an important factor in every reverse supply chain. Infrastructures are also needed in processing and recycling units. A collective alternative management system for WEEE, like any such system, is necessary to cover geographically the entire territory of a country, including remote areas. Also, the development of any such system should be based on the upgrading and extension of existing infrastructures.

The alternative management system for EEE products at the end of their useful life should be able to manage all categories of EEE products: Large household appliances, small household appliances, computer and telecommunications equipment, consumer goods, lighting items, electric and electronic tools, games, equipment for entertainment facilities and sport, medical appliances, monitoring and control devices, and automatic distributing devices. It is noteworthy that the first four categories add up to 90% of the total WEEE stream, with only large household appliances accounting for more than 40% of this stream.

Case study: excavation, construction and demolition waste management

The waste streams

The construction sector is considered to be one of the most important sectors in the global economy. The large number of technical projects executed all over the world in a continuous basis are of important economic, technical and environmental interest, since through technical projects large amounts of solid waste are produced.

Excavation, construction and demolition waste management (ECDW) constitutes one of the largest waste streams worldwide, along with mining and agricultural activities. Their quantity is estimated to account for the 25% of the total solid waste.[47]

Main sources of origin of the ECDW are the following (Figure 8.10):

- New construction works.
- Works of aesthetic and functional upgrading of existing constructions.
- Demolition of works at the end of their useful life.
- Destructions caused by extreme weather conditions (e.g. earthquakes, floods, etc.).
- Destructions due to extreme anthropogenic causes (e.g. wars).

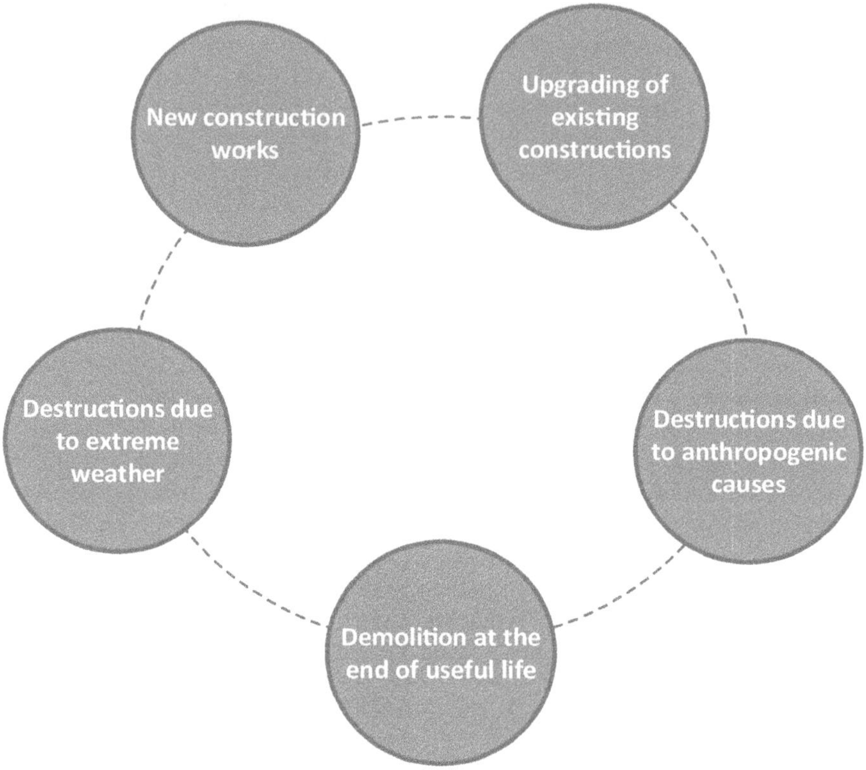

Figure 8.10 Main sources of ECDW

The type and the total quantity of the produces ECDW depends directly to the following heterogeneous factors (Figure 8.11):

- Construction type.
- Construction size.
- Building materials.
- Year of construction.
- Construction location.
- Construction, re-construction or demolition techniques.

Most of the produced ECDWs contain non-hazardous materials such as bricks, concrete, soil, stones, wood, glass, plastics, metals, packaging materials, etc. On the other hand, EDCWs may contain hazardous materials such as asbestos, lead and heavy metals, which must be separated from other ECDWs and managed in accordance with the relevant national legislation on the management of hazardous materials.

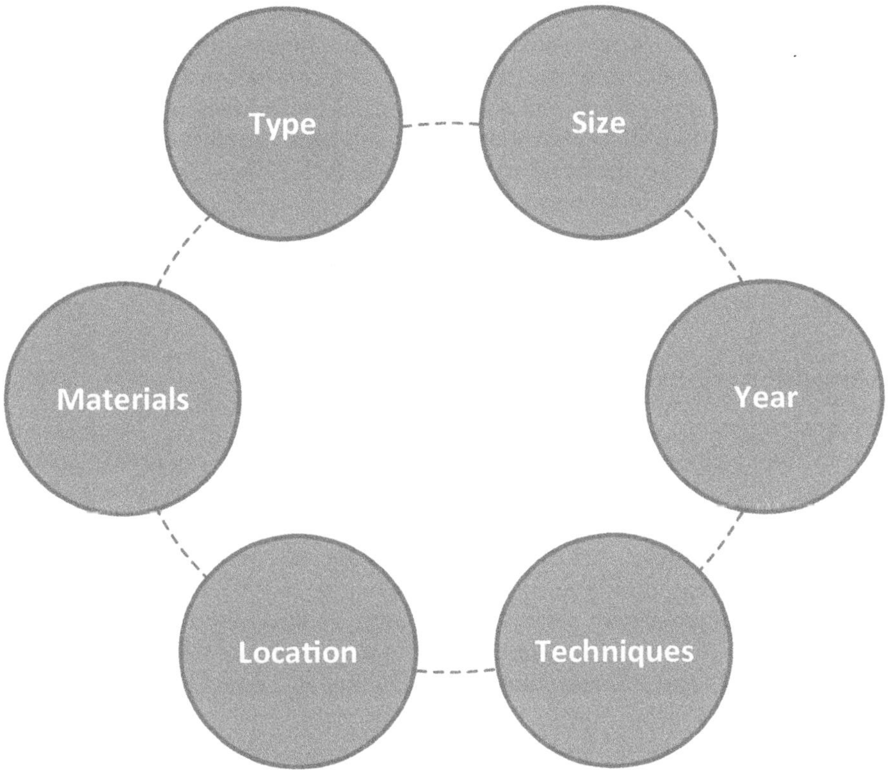

Figure 8.11 Key issues that influence the type and quantity of ECDW

The basic principle governing the commercial utilization of recovered building materials is the placing on the market of recycled materials that meet the technical specifications and have competitive prices in relation to the primary material. In any other case, the generated construction waste is driven in specially designed areas for deposition. The assessment of hazardous and toxic construction waste is of particular environmental and economic importance. The hazardous and toxic materials included in the ECDWs can be divided in the following categories:

- Materials which contain, to a lesser or greater extent, hazardous and/or toxic substances (e.g. asbestos, lead, tar, pigments and preservatives).
- Materials that are characterized as hazardous and toxic due to their stay in a contaminated environment for a long time.
- Materials that are characterized as hazardous and toxic due to their chronic contact or mixing with toxic materials during waste management.

Unfortunately, even today, in many EU member states, ECDW deposition is in illegal landfills. In many cases, illegal landfills are widespread and difficult to control, despite the measures taken to reduce their number. The specially designed ECDW deposit sites should not be confused with uncontrolled waste disposal sites, which are the sources of environmental pollution and may cause ignition. ECDWs are deposited in illegal landfills mainly for two reasons: (a) the deposition in uncontrolled landfills is considered as the most economic construction waste management policy compared to the rest of the choices; and (b) construction waste materials have relatively small resale value, since they are mostly made of inert materials, making reuse and recycling processes not economically desirable.

Dimensions of ECDW management

Recently in many countries, however, a turn towards proper construction waste management through their recycling or reuse is observed, mainly because of increased environmental awareness from the citizens' side. This is today one of the biggest challenges that modern societies have to face. One can easily notice that due primarily to the nature of the ECDWs, but also to the increased volume of the specific waste stream, the application of appropriate management policy, aiming at reuse and recycling of construction materials, is a priority with important economic, environmental, technological and legal dimensions that are further discussed in the following sections.

The recovery rate of building materials at the end of their useful life cannot be 100% in any waste management system. The same applies to the recovery of the waste generated during the useful life of the building, the management of which has a greater degree of difficulty when produced during the demolition phase compared to the production of waste at the construction stage due to significant differences in quantity and composition.

Economic dimensions

In modern times, intense mobility in the field of environmental protection in the context of sustainable development has created the need to review critical processes in many areas of economic activity, including the construction sector. The review of the critical processes related to the management policies of the produced building materials in the technical projects has very important economic dimensions, playing a decisive role in the construction sector.

It is easy to see that, in general, the implementation of sound policies for the management of ECDWs with a view to the reuse and recycling of waste increases the complexity of the supply chain processes, resulting in an increase in the overall cost of the work. In particular, a significant increase in total cost is observed at the demolition stage of end-of-life constructions, as the execution of the specific works results in the production of particularly large quantities of construction waste.

On the other hand, reuse and recycling of construction waste provides construction companies and the entire society with significant financial incentive. Recovered building materials, both from the construction and demolition stages, can be reused in other technical projects or bring revenue from their resale to recycling plants. This fact, in combination with the minimization of the total deposition cost in specially designed ECDW disposal sites, is a powerful financial incentive for construction companies. Also, by reusing and recycling construction waste, the costs for their management in the next levels of the supply chain is reduced, increasing in this way the value of the recovered materials. Additionally, from the construction companies' side, the adoption of contemporary innovative and environmentally friendly management policies can bring significant economic benefits, through the acquisition of good reputation and the improvement of their overall image on the market.

Finally, the contribution of the application of good ECDW management policies is important for the wider community through the development of the local economies. Reuse and recycling of construction materials requires the involvement of a significant number of businesses and the development of new markets for these materials, thereby creating new jobs.

Environmental dimensions

For many years, the construction sector was characterized as an environmental non-friendly activity sector, because of the large waste stream that was generated in all the stages of the supply chain (construction of new technical projects, demolition of constructions in the end of their useful life).

Significant environmental impacts are caused through the disposal of construction waste in waste disposal sites and in uncontrolled landfills. Results of the specific ECDW management practices are, except the aesthetic environmental degradation, the pollution of soil, groundwater and the atmosphere. Also, there is a constant need for new sites for waste disposal.

In this context, the application of environmentally friendly procedures in the construction sector regarding the assessment of the produces waste is a very important issue. The primary goal of the application of appropriate ECDW management policies is to contribute significantly to the prevention of waste production. This application is an economically and environmentally effective solution, since it achieves the decrease of the total construction waste management cost, while at the same time reduces the waste that ends up in landfills or in illegal landfills. It also achieves energy and water conservation.

Where waste is generated, a suitable solution is to restore the recovered material to the market either through its re-use or through recycling processes. Particular importance is given to the immediate re-use of materials, through which materials retain a significant part of their economic value.

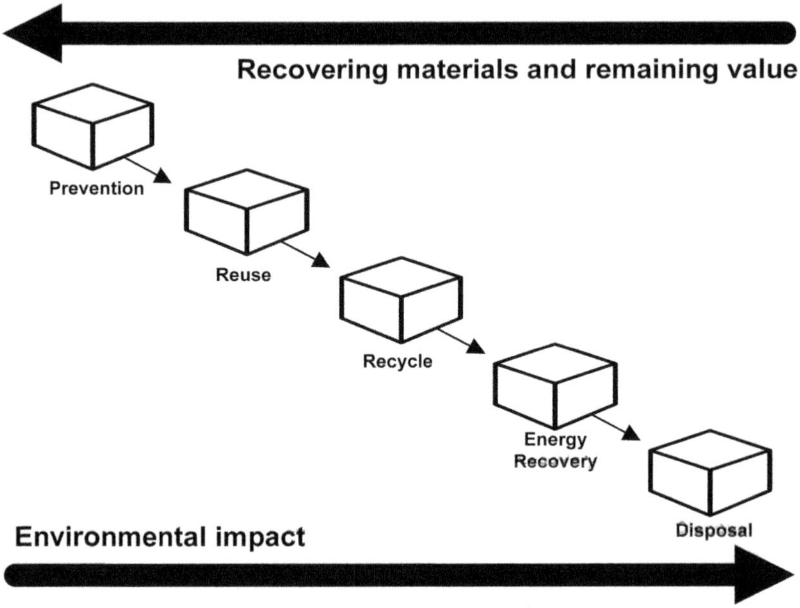

Figure 8.12 Ranking of ECDW management policies

Construction materials which are not suitable for reuse may in many cases be recycled. With recycling, materials from waste are recovered, which is the heart of every sustainable policy. The most sustainable form of recycling transforms waste into new products. Finding alternative uses for waste is another form of recycling.

If none of these methods are feasible, the waste is necessarily driven to a heat treatment or to specially designed areas where it is eventually deposited. This detailed description of the ranking of alternative ECDW management policies is presented in Figure 8.12.

Technological dimensions

The modern era is characterized by rapid technological developments in all socio-economic activity sectors. Similar is the situation in the construction sector, where technology plays a vital role with economic as well as environmental extensions, by facilitating the implementation of environmentally friendly waste management policies.

Technological advancement in ECDW management consists mainly of the parameters depicted in Figure 8.13 and analyzed in the following paragraphs.

Development and application of work execution methods for the ECDW management with the use of appropriate equipment in all of

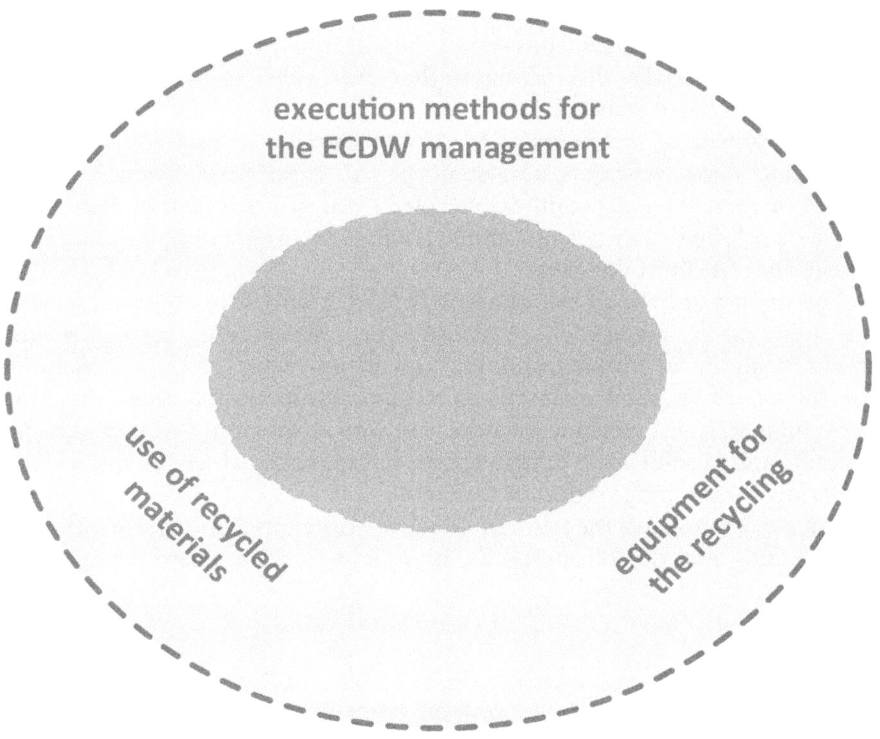

Figure 8.13 Key technological issues in ECDW management

a technical project's life cycle: The life cycle of technical projects starts from the planning phase, followed by the phase of construction of new projects and the phase of renovation of existing constructions, and is completed with the phase of demolition of technical works at the end of their useful life. Through contemporary assessment methods and the appropriate technological equipment construction companies can reduce the total cost work execution by reducing the completion time, while at the same time they are provided with the possibility to recover larger quantities of materials for reuse and appropriately assess the dangerous materials.

Use of appropriate equipment for the recycling of the produced construction materials: This equipment is basically used for the recycling of ECDW fractions from concrete, bricks, ceramics and other inert materials, which are the largest percentage of building waste. The available recycling technologies are practically divided in three levels. The first level consists mainly of the use of small capacity portable recycling units in the construction sites for direct recycling of inert materials. The second level consists of larger capacity portable recycling units, which can be used either in the construction site for the direct recycling of the inert materials or in a fixed base in specially designed waste disposal sites. Finally, the third level refers to large-scale inert materials recycling units that are placed in specially designed construction waste disposal sites.

Development of technologies for the proper use of recycled structural materials: In modern times, almost all the structural waste produced can be reused or recycled (e.g. products extracted from the treatment of inert waste can be exploited with corresponding savings in raw materials in different categories of technical works).

The main products of recycling of construction waste (skirting, general embankment fill materials) are already successfully being used in several applications in advanced technology countries. Most of the applications are the use of recycled materials as road construction materials, embankment fill materials in technical works, restoration materials for uncontrolled landfills and landfill sites, drainage material in technical works and as inert materials for the preparation of concrete

Table 8.2 illustrates the main building waste generated during a buildings construction and demolition phase, as well as the alternative management options.

Table 8.2 Construction waste management options

Material	*Reuse*	*Recycle*	*Disposal/ Specialized treatment*	*Assessment options*
Brick	X	X		Reuse of certain types/recycling as inert material
Concrete		X		Recycling as inert material
Cement tiles		X		Recycling as inert material
Concrete additives			X	Removal for specialized treatment
Tile	X		X	Reuse of certain types/recycling as inert material
Ferrous metals		X		Recycling for steel production
Non-ferrous metals	X	X		Reuse as construction materials, e.g. windows, doors/recycling for steel production
Asphalt		X		Recycling for asphalt production
Wood (without process)	X	X		Reuse/burning
Processed wood			X	Burning
Glass		X		Recycling as inert material
Ceramic		X		Recycling as inert material
Asbestos			X	Removal under controlled conditions for special treatment

Material	*Reuse*	*Recycle*	*Disposal/ Specialized treatment*	*Assessment options*
Lead			X	Removal under controlled conditions for special treatment
Plastic		X		Recycling for plastic production
Adhesives			X	Removal for special treatment
Paint and yopping layers			X	Removal for special treatment
Mineral fiber	X		X	Reuse of mineral fiber tile elements/ removal for special treatment
Gypsum board (dry construction)			X	Removal for special treatment
Porcelain	X	X		Reuse of certain types/recycling as inert material
Ventilation/fire protection systems			X	Removal for special treatment
Soil /stones		X		Use as embankment fill materials
Mixed waste		X	X	Mechanical and manual separation of materials prior to recycling as inert materials

Legislative dimensions

In dealing with the disposal of construction waste in non-controlled landfills, the European Union, the United States and many other countries have developed specific legislative actions. In advanced economies, ECDWs have been recognized as a "priority waste stream" and member states have adopted legislation aimed at the protection of the environment.

Decision-making methodological framework for construction waste management

Figure 8.14 presents the material and information flow of the construction industry supply chain. The traditional supply channel consists of the stages of raw material extraction, the production and supply of construction materials and finally the building construction stage. Then the building's use stage follows, which includes any maintenance or renovation works that may take place during the building's useful life. The reverse supply chain mainly concerns the end of the building's useful life – in particular, its demolition stage. The main characteristic of the particular supply chain is the fact that, throughout the production, significant quantities of construction waste are produced, especially at the demolition stage of the building.

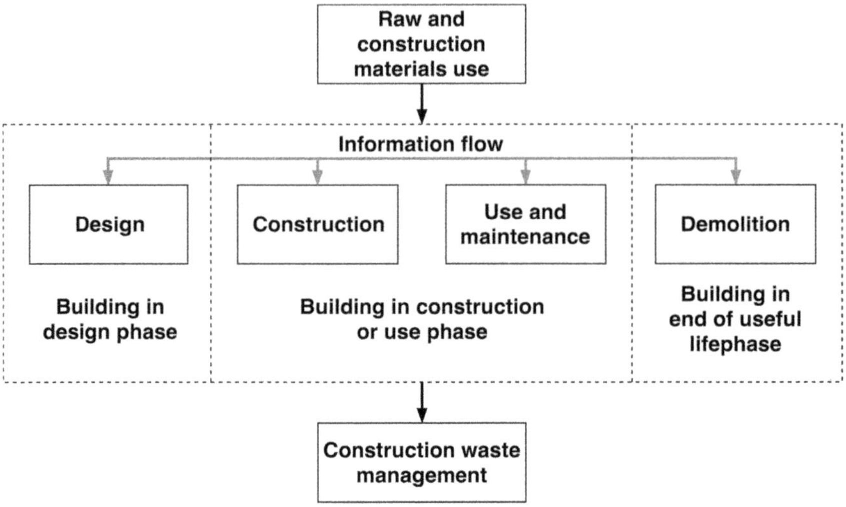

Figure 8.14 Material and information flow in the construction industry supply chain

Design phase

The design phase is a very important factor in the assessment of waste produced by construction works. According to Innes,[48] it is estimated that the 33% of construction waste in construction worksites is caused by the failure to implement appropriate waste reduction measures during the design phase. The main design factors that lead to the increase of construction waste are the following:[49]

- Change in design during the work phase.
- Complexity of construction details.
- Design using low-quality building materials.
- Lack of designer experience of construction methods.
- Lack of designer knowledge of the existence of alternative construction materials.
- Design using materials that do not have standard dimensions.
- Unclear specifications of materials and jobs.
- Lack of coordination between the developer, the designer and the manufacturer.

From this list, it is easily perceived that the development of an integrated waste management plan that includes the design stage can positively contribute to the reduction of the waste produced. The basic principle of designing new construction is the flexibility of design for any changes that occur during the construction phase to incorporate the changes without the production of waste. In addition, other basic principles of new constructions design are summarized here:

- Estimation of the q\uantities of waste generated.
- Analysis of the use of specific construction materials.
- Redesign of the whole construction process aiming at the use of alternative materials, reducing, in this way, the production of waste.
- Minimization of changes in design during the construction phase.
- Reduction of the total use of construction elements and materials.
- Detailed explanation of design details.
- Use of high quality materials.
- Use of materials with standard dimensions and formatting of designs according to standard material dimensions.
- Avoiding use of hazardous or toxic construction materials.
- Avoiding use of materials that are sensitive to wear and contamination
- Use of materials that are easily recycled.

Another important decision in the design of new constructions is related to material management at the end of the construction's useful life.

Designing of new construction aiming at the optimal material management at the end of a building's useful life is an emerging architectural concept that uses basic principles from fields related to the consumer products industry, in particular with the design for dismantling, reuse, rebuilding and recycling of consumer products. The main objective of this sector is to respond to any future changes as well as to reduce the waste generated by increasing the rate of reuse or recycling of building elements and materials at the end of the useful life of the construction.

Taking into consideration that the different parts of a construction may have a different life cycle, due to the use of many heterogeneous construction elements and materials with different useful lifespan, the new construction design that aims at the optimal assessment of materials in the end of the buildings useful life is based in four axes (Figure 8.15):

- Design considering future change of use of the building, in constructions with a long life expectancy
- Design based on relocation of the entire construction.
- Design based on the reuse of construction elements of the structure after its selective deconstruction.
- Design based on the reuse or the recycling of certain materials after its selective deconstruction.

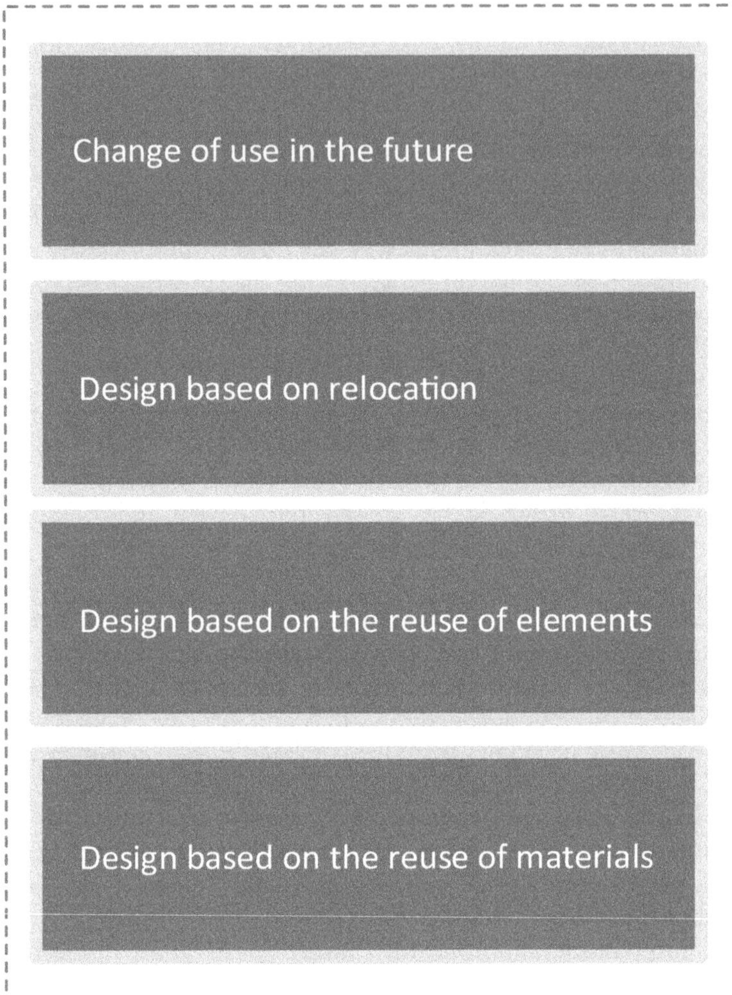

Figure 8.15 Parameters that need to be considered in the design of a building

The basic principles of designing new constructions for optimal material management at the end of the useful life of the building are summarized here:

- Use of high-quality reused and/or recycled materials, through which local market will be encouraged to recover construction materials.

- Use of materials that are produced by the local market for direct reuse locally at the end of the construction useful life.
- Use of materials that have low weight and specific dimensions that facilitate their handling during the selective deconstruction.
- Placement of construction elements and materials in order to facilitate parallel works during the selective deconstruction of the construction.
- Avoid use of hazardous or even toxic materials, which are difficult to handle at the end of the useful life of the construction.
- Reduction of the total use of construction elements and materials, which will result in the reduction of the waste generated during the selective deconstruction of the building.
- Reduction of the use of different materials, contributing to the simplification of the selective deconstruction process.
- Facilitation of the construction materials' reverse flows at the end of the construction's useful life mainly by providing accessibility at all the construction stages.
- Facilitation of future renovations.
- Reduction of the total connection number, through which the selective deconstruction processes are simplified.
- Simplification and establishment of specific mechanical connections (where possible), contributing to the simplification of the selective deconstruction processes.
- Keeping all the information on the materials used, to develop an appropriate management plan for ECDWs.

To sum up, it is easy to understand that the design of new constructions aiming at the optimal management of waste generated during the construction phase as well as the design of new structures with the aim of optimal material management at the end of the building's useful life are governed by common basic principles. Therefore, designing is a critical stage of the whole process, in which appropriate decisions that relate to the entire life of the building should be taken, from the structure's construction to its demolition.

Construction phase

Waste produced during construction works are mainly material wastage. The main waste produced at this phase are concrete, wood, metal and plastic materials. The production of waste at the construction phase takes a significant percentage of the total weight of the construction materials delivered at the construction site. According to Cheung et al.,[50] this percentage can be up to 10–20% of the materials' total weight. It should be mentioned that in many cases it is difficult to determine the exact composition of the waste produced, due to the alternative construction techniques adopted by project managers. The main factors at the construction phase that result in the increase of the production of construction waste are:

- Waste production due to negligence or workforce mistake.
- Lack of workforce knowledge for the proper use of the material and the equipment used during construction works.
- Damage caused by malfunction of the equipment.
- Use of inappropriate material, according to specifications set at the design phase.
- Inappropriate storage of material at the construction site.
- Mistakes at the material procurement (order size, type of material, material quality, supplier errors, etc.).
- Damage during transportation of the materials from the supplier to the site or from the temporary storage area to the site.
- Inability of the project manager to perform supplies of materials in small quantities.
- Theft of materials at the site and damage caused by vandalism.
- Wear of materials due to weather conditions and improper temporary storage at the site.

The minimization of the produced waste from the construction phase can be achieved through the development and application of a management plan. The management plan is an integral part of construction works and is directly relevant to all the participants in the supply chain (suppliers, project manager, construction workers, and recycling companies). At this stage, the most important decisions and measures included in the management plan are summarized here:

- Predetermination of the type, the total quantity and the management method of the produced construction waste. Emphasis should be placed on the management of hazardous materials.
- Collaboration across the supply chain (suppliers, project manager, construction workers, recyclers of construction waste).
- Proper education of the personnel involved in the construction phase. Education aims at informing the personnel about waste minimization techniques through good working practices, as well as provide with information regarding the possibilities of the collection and management of the waste produced.
- Selection and application of construction techniques for the minimization of the production of construction waste.
- Onsite reuse of materials.
- Recycling of construction waste produced during construction phase.
- Avoid using, where it is possible, toxic and hazardous construction materials.

- Application of the "just-in-time" policy for the procurement of materials, with the aim of maintaining the minimum possible inventory at the construction site and receiving materials at the time they are needed to handle specific construction work. In this way, it is possible to avoid damages of the materials caused during their storage on the construction site and due to their additional movements within the worksite.
- Supply and use of recycled construction materials.
- Supply and use of high quality materials in accordance with design specifications to avoid future replacement of materials.
- Supply of materials in the appropriate dimensions, in accordance with the design specifications, to prevent wastage.
- Qualitative and quantitative control on receipt of materials for the immediate identification of damage caused by transport or due to inadequate packaging.
- Continuous inventory control of building materials to avoid surplus supply.
- Selection of a central point for the temporary storage of materials within the worksite in order to minimize movement.
- Provision and use of the necessary special bins for the temporary storage and transport of the waste generated by species to the final recipients.
- Continuous control of the construction waste produced at the site and redefinition of the waste management plan.

Demolition phase

Demolition is the final stage of the building's useful life following the designing, construction and the use and maintenance phases.

In some cases, the procedures for the demolition of a building can be considered as the reverse to those of its construction, but it is virtually impossible to adopt the same strategies and methods for managing the waste generated. The most important reason is the type and the total quantity of waste generated, as in the case of the construction of a new building the waste generated is a byproduct of the main activity, while in the case of demolition, waste is the main product of the entire process.

In recent years, innovative methods and strategies have been developed for the sound and efficient management of building demolition and waste generated. The management strategies concern both the choice of the best demolition technique and the optimal way of managing the construction waste generated.

The objective of the optimal management of the produced construction waste from demolition, can be achieved through the development of an integrated methodological framework for the management of construction waste. The proposed framework consists of two phases:

- ***Phase I:*** Evaluation of construction and optimization of the procedures for performing tasks and recovering materials.
- ***Phase II:*** Implementation of the procedures for performing tasks and recovering materials.

During the initial Phase I stage, the following tasks should be performed:

- ***Building Assessment:*** Key features to be considered are the date of the building's construction, the overall size and height of the structure, the location and the various characteristics of the area and the construction method followed, since it greatly affects the recovery of materials.
- ***Assessment of Building materials:*** Analytical assessment of the different materials, their condition and estimation of the total amount of waste generated by creating a bill of materials. It is also particularly important to record the possibilities of re-use and recycling of building materials that compose the construction.
- ***Environmental assessment:*** The basic task prior to demolition is the identification of the hazardous and toxic materials located in the building. The existence of these materials requires special handling for their removal and their final disposal.

In the second stage of the initial evaluation, the licensing of the procedures for the demolition is reviewed, the approval of which is required by the competent authority for any demolition of a building.

The next step concerns the overall planning and organization of the procedures that will take place for the implementation of the project. Critical decisions relate to choosing the best demolition process and determining the depth of application of selective deconstruction. In other words, the project manager must decide the type and quantity of materials to recover before carrying out the conventional demolition works on the remainder of the construction. Additionally, decisions should be made on the management processes of the produced construction materials. The final decisions of this stage concern the design of an optimal network for the transport of materials from the construction site to the final recipients (specially designed ECDW deposit sites and construction material recycling companies).

In the final stage of Phase I, a construction site check is carried out to ensure the execution of the proposed demolition works. If any risks to the safety of workers are identified, it is mandatory to redesign all the procedures and take the necessary safety measures before the operations begin.

In case the proposed solution is economically, environmentally, legally and technologically reasonable, then the project manager proceeds to Phase II of the framework. Preparing the worksite is the first stage of this phase.

The site must be properly prepared for receiving the equipment and the special waste bins for the construction waste produced. Subsequently, demolition works are carried out. The last step concerns the procedures for the recovery of construction materials and their transfer to final recipients.

Alternative demolition procedures

The contribution of initial assessment to decision-making on the demolition process is important, especially when choosing the technique to be adopted. Today, alternative demolition techniques have been developed due to the rapid development of technology and equipment in the construction industry. In particular, two are the most basic techniques applied for the demolition of buildings, selective deconstruction and conventional demolition.

Conventional demolition

Conventional demolition involves the execution of works aiming at the permanent demolition of the building in the short time using mechanical means such as hydraulic crushers, excavators, scissors and hammers. In some cases, explosives are used, especially in the case of high building demolition.

The most common method of conventional demolition is referred to in the literature as the "top-down" method. As indicated by the name of the method, the demolition takes place on a vertical axis with downward execution. In other words, the demolition begins with the dismantling of the roof and is completed on the ground, demolishing the building floor to floor. According to Kourmpanis et al.,[51] the "top-down" method is ideal for the demolition of buildings located in densely populated areas where access to heavy machinery is difficult. This method has several similarities in its application with the selective deconstruction method.

Conventional building demolition results in the creation of mixed waste, as it is often impossible to separate the produced materials and re-use or recycle them. In rare cases, a rough manual separation of waste occurs at the point of production (site) after the demolition of the building. The application of conventional demolition is appropriate when there is no time and financial flexibility.

Selective deconstruction

Selective deconstruction can be defined as the method of demolition of a building, in a way that ensures the safe removal and disposal of hazardous materials and maximize reuse and recycling of the remaining materials. The process of selective deconstruction of the building can be characterized as the reverse of its construction process.[52] In other words, the final material placed during the construction of the building is first material that is removed when carrying out the operations of selective deconstruction.

The method of selective deconstruction attempts to maximize the recovery of materials within a certain time frame at the lowest possible cost. These materials should be removed either because of their economic value or historical importance, or because they can be readily reused without further treatment or because their presence pollutes or reduces the quality of the waste stream (e.g. the presence of hazardous or inorganic components in the inert waste stream).

On the other hand, selective deconstruction is a work-intensive process as the removal of the various materials from the building is done manually to ensure the high quality of the materials produced.

The process of selective deconstruction consists of distinct stages, as at each stage different materials are removed from the building and collected in special bins for re-use or recycling. The first stage involves the removal and collection of hazardous and toxic materials, e.g. asbestos, lead and others. In the next stages, materials that can be reused – such as doors, windows, floors and electrical and plumbing parts – are removed. The final stage is the conventional demolition of the building frame using the appropriate mechanical equipment.

Comparison of alternative demolition processes

The comparison of alternative demolition processes is outlined in Table 8.3, which presents the advantages of carrying out selective deconstruction and conventional demolition work.

From this analysis, it is easy to understand that the implementation of selective deconstruction has more environmental and technological benefits compared to conventional building demolition. In particular, the management of the waste generated in the context of the selective deconstruction process makes this technique environmentally friendly, while saving valuable resources. According to the US Environmental Protection Agency (EPA), "the selective deconstruction technique can be the source of tackling the problem of global warming by reducing solid waste production".[53]

On the other hand, there are significant limitations to the application of selective deconstruction as a demolition process, which stem mainly from the fact that selective deconstruction is a relatively recent technique and that the necessary supporting functions, markets and regulations have not been fully developed to date. Table 8.4, presents the main restrictions on the application of selective deconstruction.

Chapter summary

Solid waste is considered to be one of the main pressures on environment. The volume of the waste generated is constantly growing due to the increase of the development rate, but also due to change of consumer patterns. For the management of waste, the following alternatives should be co-exploited

Table 8.3 Comparative presentation of the advantages of carrying out selective deconstruction works and conventional building demolition

Advantages	*Selective deconstruction*	*Conventional demolition*
Environmental	• Elimination of environmental impacts (use of hand tools and small equipment) • Low probability of material contamination (separation of toxic materials from the rest) • Extension of the materials' life cycles • Small quantities of materials to be deposited • Reduce energy use	
Economic	• Income from the sale of recovered materials • Low waste disposal costs • Low equipment cost • Low cost for materials management at the next stage of the chain • Economic growth (multiple enterprises involved) • Creating new jobs	• Low total operational costs • Low labor costs
Technological	• Appropriate hazardous waste management • Recovering architectural elements and materials of historical value • Higher recovery of materials for re-use • Employee opportunity for continuing education	• Ability to apply alternative techniques • Smaller task duration • Use of a small number of special bins

Table 8.4 Major restrictions on the application of selective deconstruction process

Restrictions in selective deconstruction
• The majority of structures, or parts of them, have not been designed with respect to selective deconstruction • Economic and environmental benefits resulting from the recovery of materials have yet to be introduced • Lack of appropriate market for recovered materials • High implementation costs • Lack of suitable mechanical equipment to remove specific materials from the building • Absence of technical and quality standards for recovered materials • Low cost of material disposal in specially designed sites • Use of a large number of special bins for material storage • Non-competitive prices of recovered materials • Increased work completion time • Increased labor costs • Difficulty in establishing a market for recovered materials (high degree of uncertainty for the disposal of recovered materials)

in an integrated manner – minimization of waste production, reuse, recycling and energy recovery. In this chapter, we focused on the cases of reverse logistics for waste electrical and electronic equipment (WEEE) and excavation, construction and demolition waste (ECDW), which are two significant waste streams due to their hazardousness and volumes. Following an introduction to the specific waste streams, available alternatives were discussed, and integrated approaches for their management were promoted.

9 Decision-making in the green supply chain

Discussion questions

- What are the key assessment frameworks and methods in green supply chain management?
- What are the key assessment indicators in green supply chain management?
- What are their key benefits for the decision-maker?

Introduction to assessment methods

In order to achieve sustainable development goals, it is essential to assess the entire process through examination and keep track of the progress made. The application of assessment methods is important in order to estimate and assess the improvements and changes that are made towards implemented schemes and processes. Assessment methods are developed to assist decision-makers with pending decisions by providing an overview of the capacity of the environment while at the same time evaluating the effectiveness and the efficiency of the examined subjects and processes.[54] Assessment methods can be considered as effective policy making and communication tools since they provide policy makers with the relevant information to model the examined issues. The application of such methods can clarify, measure, analyze, determine and communicate multidimensional and complicated issues.

Despite the widespread application and use of assessment methods, to this day there is no consensus in the scientific community on the common framework under which they should be put.[55,56] As a matter of fact, scientists are divided into two categories. On one hand are those who support the statement that assessment methods do not actually add much to sustainable development since they are unable to assess processes and services and only focus on market goods.[57] On the other hand, there are supporters who believe that assessment methods contribute to the promotion of sustainable development.[58–60]

Assessment methods and frameworks such as cost-benefit analysis, contingent valuation, hedonic pricing method, travel-cost method, multi-criteria analysis as well as statutory tools like environmental impact assessment (EIA) and strategic environmental analysis (SEA) were introduced long before the widely known Brundtland Commission in 1987, a commission established by the United Nations in order to promote sustainable development globally. However, following the publication of the Brundtland commission's report (i.e. "Our Common Future"),[61] a large number of assessment methods and tools emerged. To date, there are numerous assessment methods which would be difficult to categorize since they employ a variety of methodologies and have different characteristics depending on the subject under examination. In this section, an overview (Figure 9.1) of the most widespread and widely accepted methods is presented.

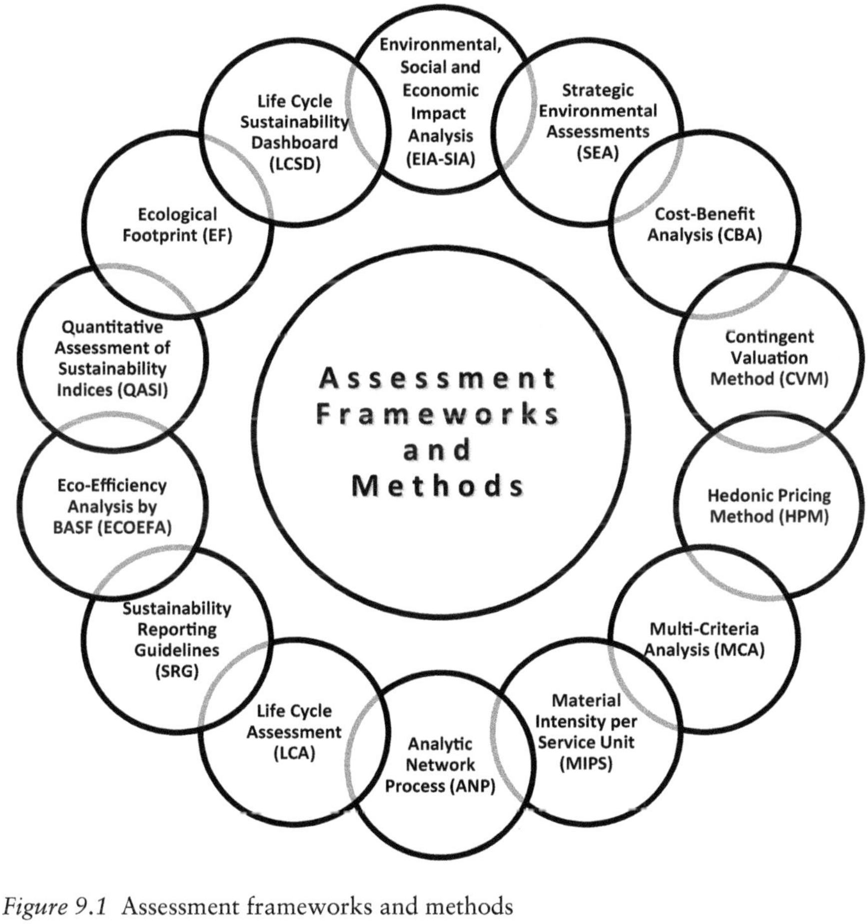

Figure 9.1 Assessment frameworks and methods

Frameworks and methods

Environmental, social and economic impact analysis

In 1969, under the US National Environmental Policy Act (NEPA), the concept of environmental impact analysis (EIA) was introduced. Impact analysis aims at assisting decision-makers as well as the stakeholders by providing information about the impacts of a project or activity prior to its realization. The EIA process takes into account the environmental impacts of the proposed projects and activities. Social and economic impact analysis considers the relevant impacts and along with EIA provide decision-makers with integrated information on the sustainability performance of proposed projects and activities. Despite the facilitation of the assessment of various impacts, the process has been said to have some insufficiencies in the field of impact prediction, definition and measurement, impact monitoring, use of method, consultation and participation.[62]

Strategic environmental assessments (SEA)

As presented in the previous section, one of the disadvantages of EIA is that it is applied only to a certain project aiming at determining the impacts of its application. To address this insufficiency, the United Nations Economic Commission for Europe suggested the expansion of EIA in order to assess policies, plans and programs (PPP). The outcome of this initiative was the introduction of strategic environmental assessments (SEA). SEA assists decision-makers by safeguarding that the relevant information is available in order to make appropriate and on-time decisions. SEA can be applied in higher levels on the decision-making process in contradiction with EIA. One of its main advantages is the promotion of public participation, however as a process, it can be characterized as costly and resource consuming. It is worth mentioning that SEA usually does not take into account social and economic impacts. Also, SEA has issues regarding the structure of public participation, data availability and the assessment of uncertainties.[63]

Cost-benefit analysis (CBA)

The cost-benefit analysis (CBA) method considers the costs and benefits of the activity or task under examination. CBA is a decision-making tool which, when used at the initial steps of a project or activity, can provide with insights regarding its feasibility. The method can be used by decision-makers to examine alternatives in the implementation of a project or compare the costs and benefits of a series of different activities. The most widespread tools, related to capital budgeting, that are used in CBA are the internal rate of return (IRR) and the net present value (NPV). The NPV is the process of adjusting all future cost and benefits by transforming them to their present values, while the IR R is the discount rate that nullifies the net present value of an investment.

When conducting an economic CBA, all the costs considered are the expenses of the development, given in financial terms and reduced to the time value of money.[64] On the other hand, the benefits are the profit generated from the operation of the development. Cost-benefit analysis is a methodological approach to assess different developments or activities by comparing them based on their present value. The CBA method has some deficiencies which are attempted to be considered with the introduction of akin tools like cost-effectiveness analysis, community impact analysis, cost-utility analysis, economic impact analysis, social return on investment, and analysis and fiscal impact analysis.[63]

Contingent valuation method (CVM)

The contingent valuation method (CVM), which is also described as the "stated preference" method, is used to evaluate the economic value environmental related projects and activities. It can be used to estimate the willingness to pay for environmental developments, and also the willingness to accept of activities or projects that result in environmental degradation.[65] It is based on surveys in order to obtain the relevant information, and to that end the questionnaire design is of utmost importance for the application of the method along with the hypothetical plan determination. People involved in the surveys conducted, should be knowledgeable of the hypothetical plans especially since some of them may need certain expertise in the respective fields. One of the main advantages of CVM is that it is adaptable and can be used to monetize and evaluate non-market goods. However, the boundaries of the method's applicability are limited and cannot be expanded to whole ecosystems.

Hedonic pricing method (HPM)

Hedonic pricing method (HPM) uses regression analysis to infer environmental values from spending on goods that include those values. The HPM was introduced by Rosen[66] and is founded mainly on Lancaster's theory. The method assesses the economic impact of environmental services to the market. The ultimate goal of the method is to establish the connection between the value and the characteristics of a product or service. For example, a product resulting from a green supply chain may have a higher market value.

Multi-criteria analysis (MCA)

Multi-criteria analysis (MCA) is a widely known decision-making tool designed to assess controverting alternatives when complicated frameworks and values have to be considered.[67] The MCA tool assigns weighting factors in order to evaluate impacts in non-financial terms. This is one of its main advantages, since it provides with the ability to assess alternatives against multiple criteria measured in different units. It is also flexible with the judgment of the decision-making team. One of the main disadvantages of the method is that it is susceptible to uncertainties. Also, the inputs of the mathematical model strongly depend on the individual assumptions of the experts involved.

Material intensity per service unit (MIPS)

Material input and output flows are part of every product or service. A framework proposed to record and evaluate these flows is the material input per service unit (MIPS) and the similar ecological rucksack, both developed by Wuppertal Institute. In MIPS, a product is analyzed into material flows that were necessary for the product to be manufactured. The total material intensity (MI) is divided by the number of service units (S). The method resembles the ecological footprint methodology, but is industry-focused and directly applicable to production or supply chain systems. The environmental impact is quantified and includes all material flows for the life cycle of the examined system. The ecological rucksack differs slightly by subtracting from the inputs the weight of the final product itself. Conversion factors to estimate MI for various inputs exist.[68]

Analytic network process (ANP)

Analytic network process (ANP) is a multi-criteria analysis tool, similar to analytic hierarchy process (AHP). While AHP is structured as a sequential linear system consisting of goal, decision criteria and alternatives, ANP presents the decision problem as a network allowing for interrelationships to be recognized and documented. ANP components include elements that form clusters, control criteria relevant to the elements and all interdependencies between criteria, elements or clusters. ANP entails the following steps:

1) Structuring the decision problem, listing all elements, criteria, actors, objectives, etc.
2) Paired comparisons for criteria, elements and clusters. These steps output priorities to be used as input for the next level paired comparison.
3) Sensitivity analysis of the final weightings.

Non-linear approaches as ANP offer more flexibility and allow complex systems to be modeled with less ambiguity and conditional assumptions.

Life cycle assessment (LCA)

Life cycle assessment (LCA) determines the environmental impacts of a product or service throughout its life cycle. LCA is a great tool for decision-makers in order to improve the environmental performance of a product or service from "cradle to grave". LCA considers the impacts all the stages in the life cycle of a product or service from the material extraction, manufacture, packaging and transport, use and its final disposal. LCA as a tool is standardized in ISO (14040 series). The application of LCA is impeded by data requirements (LCA strongly depends on data availability), methodological discrepancies and technical attributes. LCA contributes to triple bottom reporting by quantifying the environmental performance of a project or activity.

Sustainability reporting guidelines (SRG)

A structured reporting method proposed by Global Reporting Initiative, an independent international organization that specializes in sustainability reporting, is the sustainability reporting guidelines (GRI) framework. The framework responds to the need of a standardized reporting scheme that ensures quality and applicability of sustainability reports. The method can be modified by compiling a list of fitting indicators for each activity sector examined. Indicators used are numerous, more than 100 in the latest review, and cover a wide array of activities. According to the procedure, 17 core indicators and 13 supplementary indicators are used and depending on their assessment and their consistency check, internal or external, a sustainability level is reported on an A–C scale.

Eco-efficiency analysis (ECOEFA)

Eco-efficiency analysis (ECOEFA) is a method internally developed by the BASF company in 1996. Being the product of a corporation, it is innately business oriented and therefore includes the financial dimension as a critical factor along with environmental impact, a "[f]rom cradle to the grave with costs" analysis.[69] The method assigns values for financial and environmental performance to each alternative to render them comparable and plot an eco-efficiency portfolio and eco-efficiency index chart.[69] To examine environmental impact, five main factors are weighed in:

1) Raw material consumption.
2) Energy consumption.
3) Emissions production.
4) Toxicity potential.
5) Abuse and risk potential.

Methodologies to assess the impact of pollutants and the incurred costs are part of the toolkit and cover all aspects examined. Resulting data can be plotted on two diagrams. The Eco-Efficiency Portfolio is a two-axis diagram of x-values representing economic impact from higher to null and y-values representing environmental impact from higher to null. The second manner of presentation is the eco-efficiency index (EEI), which produces a single-scale diagram that combines both impact factors to rank alternatives in a single index.

Quantitative assessment of sustainability indices (QASI)

The quantitative assessment of sustainability indices (QASI) method[70] quantifies several normalized impact indices to assess sustainability of the possible alternatives. To implement the method, four key steps are required:[70]

1) Selection of reference criteria to define process alternatives.
2) Definition of quantitative indicators.
3) Normalization of indicators.
4) Aggregation of indicators to the resulting sustainability indices.

The method is designed to be applicable to early process design stages, as well.[70]

Ecological footprint (EF)

Ecological footprint (EF) is a term coined by Wackernaegel and Rees in 1996[71] presenting the link of consumption to resources through equivalent areas needed to produce the consumed resources and the waste/emissions incurred.[71] The estimation provides a bulk assessment of environmental impact, and a rough estimate of sustainability can be calculated since resources are not infinite. However, the application of EF to industrial applications is not a straightforward process; for example, global and local impacts need to be considered in relation to equivalents reported by the method.

Life cycle sustainability dashboard (LCSD)

The dashboard of sustainability provided the basis for the life cycle sustainability dashboard (LCSD)[72] which outlines necessary modification steps for the framework to serve as a sustainability benchmark utility for products. The examination period refers to the whole life cycle, in a "from cradle to grave" approach. Critical input to this method is also the selection of indicators that will be considered during the assessment. The developers of the method launched an online tool with free access that can be used to apply the method. The results are a graded performance and graphical representations that aid in communicating the findings more effectively.[72] The open architecture of the application allows it to be versatile and adaptable to virtually any scenario, by choosing the desired indicators and entering performance values. The output includes environmental, social and economic dimensions, presenting a comprehensive profile of the current status.

Assessment indicators

A structured assessment framework needs to be able to report performance in a concise and consistent way. Indicators serve this purpose, providing information and monitoring the impact of actions related to green supply chain activities. In general, indicators can be categorized according to primary function as follows:[73]

- Systematic monitoring of environmental changes.
- Early warning of environmental problems.
- Target setting.
- Performance reviews.
- Public information and communication.

Their contribution to a decision support system is described[74,75] by identifying the challenges related to:

- Structure complexity and communication of information (information-structuring challenge).
- Operationalization of sustainable development (interpretation challenge).
- Social learning (interpretation and influence challenge).
- Demonstration of accountability and benchmarking (influence challenge).
- Identification of knowledge and data gaps (information-structuring challenge).

Firms aiming to sustainably develop in the future need to operate within guidelines that favor this outcome. Information is vital to reaching useful conclusions and formulating grounded assumptions. Indicators are valuable tools in this process as they convey structured information[76–78] to organizations committed to sustainable development. Sustainability involves a wide array of subjects, making including and interpolating factors a daunting and subjective task. Indicators designed to incorporate and interpolate impact and information flow of various sectors can provide a reliable reference point for analytical examination and assessment.

Furthermore, indicators serve also as a formal standardized system that allows sustainability dimensions to be integrated into existing formal business decision support tools. Fulfilling such requirements also allows for future formulation of structured methods that optimize operations or benchmarking frameworks that underline applicability to different industries within the supply chain.

The information flow involved in recording and monitoring indicators is an established communication channel between stakeholders[79] that bridges strata gaps between them. As each group is more aware of each other's contribution and influence on the system, an opportunity to adapt mindsets to multifactored conditions arises. In the best-case scenario, indicators serve also as stimulants for innovative methods development.

As mentioned earlier, data in the form of indicators are a communicative tool. Stakeholders for environmentally responsible corporations are actors beyond the supplier-user axis. According to Bebbignton et al.,[80] sustainable indicators can be used to demonstrate accountability to society.

Another noteworthy attribute of indicators is that they provide a commonly understood language to all experts involved. By design, they are meant to effectively quantify information from multiple sources and simultaneously be universally comprehensible.

Consequently, organizations that are responsible for maintaining standards, validating function and providing roadmaps to application have been founded. An attempt to catalog and provide a self-assessment tool by the European Union was made that resulted to the European common indicators (ECI). Though originally developed for urban sustainability projects, it has an open architecture by indent to be able to incorporate other subjects too.

The World Economic Forum, Yale and Columbia University have developed two district sets of indicators, the environmental performance index (EPI)[81] and the environmental sustainability index (ESI).[82] Both indicator sets are suitable as policy maker decision support systems. The EPI is a quantified system that makes information on various sectors transparent to audiences, whether or not they are experts. ESI serves mostly as a structured method for policy makers to analyze and benchmark practices and carriers of them. Through ESI, nations can be ranked according to their sustainability rating on 21 indicators. Both processes are transparent and publicly available. In response to ESI and EPI, the European Union has developed the European benchmarking indicators (EBI) to measure the environmental performance among member states.

Other indicators also exist that offer more specialized measurements, for example indicators of sustainable production (ISP), measuring production and Wuppertal sustainability indicators (WSI), a holistic method that combines various aspects and is applicable both in corporate or nation level, similar to the organizational sustainability performance index (OSPI).

Chapter summary

In order to achieve sustainable development goals, it is essential to assess the entire processes under examination and keep track of the progress made.

The application of assessment methods is important in order to estimate and assess the improvements and changes that are made towards implemented schemes and processes. The application of such methods can clarify, measure, analyze, determine and communicate multidimensional and complicated issues. In this chapter, we focused on the key assessment frameworks and methods in green supply chain management, along with the key assessment indicators that exist in the field. The key benefits for the decision-maker from the use of a number of available tools were analyzed.

10 Technologies of marketing and supply chain management synergies

Discussion questions

- What are "green information and communication technologies", "smart ICTs" and "sustainable ICTs", and how are those related to green supply chain management?
- What areas do such technologies mostly affect in the supply chain of an enterprise?
- What types of "green information and communication technologies" exist in the market?

Introduction

Green information and communication technologies (GrICT) refer to all the technological solutions that can be used to improve environmental performance throughout the economy and society.[83–85] Overall, GrICT have better environmental performance and are referred to in the literature as "smart ICTs" or "sustainable ICTs". The correlation between green information and communication technologies (GrICT) and green supply chain management (GSCM) includes two axes of goals, as illustrated in Figure 10.1:

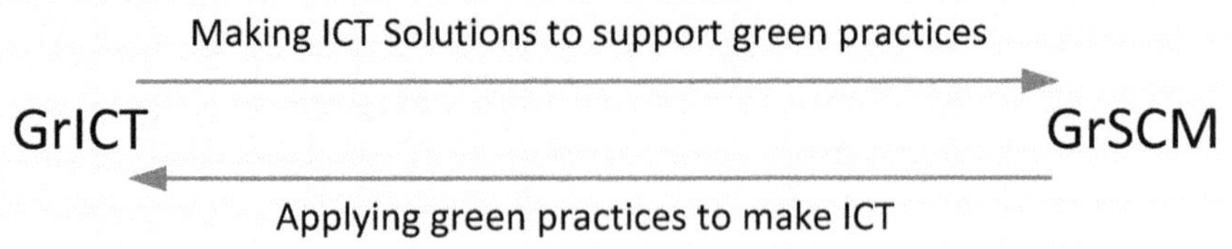

Figure 10.1 Aspects of GrICT and GSCM relations

In simple terms, the objectives are:

- ***Making ICT solutions for optimizing the application of green practices (ICT for Green):*** Namely, the implementation of technologies and information systems for the measurement, monitoring and the general management of data relating to the effects of business activities on the environment. The data gathered are then processed with one main objective in mind, to support the decision-making process that will lead to actions for the protection of societies and the environment. That's why GrICT can be considered a critical factor for the success of "green growth".
- ***Making ICT solutions using green practices (Green for ICT):*** Namely, the effective and efficient management of the activities involved in the information system development cycle and computational systems with low or at least admissible effects on the environment. Managerial effectiveness means that by performing the activities the objective is also attained, while efficiency occurs when the activities are performed in the optimal way, namely faster, cheaper, etc.

This chapter has been organized and discusses both axes, but it focuses on the first one.

Making ICT solutions for managing green practices

Pursuant to the definition offered in the first chapters, GSCM constitutes a way to manage business activities using green/environmentally friendly practices and technologies.[83] The supported business functions include the design of products and services, the selection and procurement of raw materials and semi-finished products, the process for the production of the finished products, their transportation and distribution to consumers and, finally, their management after their useful, operating or use life cycle (Figure 10.2).

ICT solutions aim for the integration and optimization of all the GSCM operations and the supporting of SC decisions on all business levels (strategic,

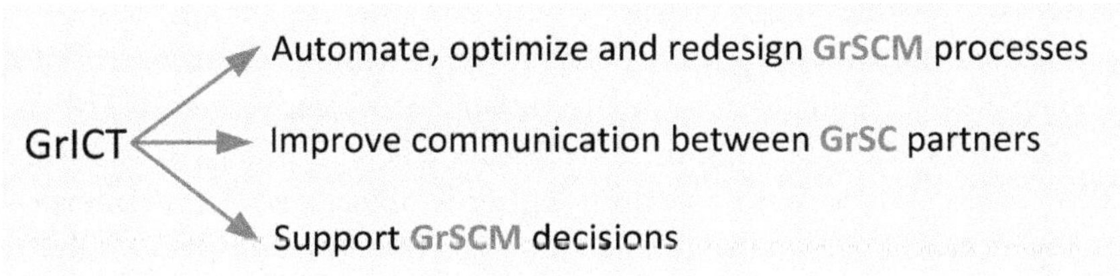

Figure 10.2 Aspects of GrICT in GSCM

tactical and operational decisions), as well as offering best green practices and methodologies.[86]

Specifically:

- GrICT solutions aim to achieve the automation, simplification, optimization and redesign of GSCM processes, such as procurement/sourcing, manufacturing (including product design by replacing hazardous materials or processes by more environmentally friendly ones or by measuring their environmental performance), re-manufacturing (including product recycling – repair – disassembly – reuse – recovery processes), warehousing (such as energy saving, green inventory management, etc.), supply chain network design (location, distribution, reverse logistics, inspection/sorting of used products, etc.) and waste management (source reduction, disposal, etc.). Furthermore, it includes the measurement of the environmental impacts of supply chain and logistics systems, which remains an important issue to address because (as was cited in previous chapters) of the many and diverse methodological approaches that are available.[87,88]
- Moreover, GrICT solutions improve the communication and achieve the coordination, cooperation and integration of supply chain partners (both suppliers/trading partners as well as customers) in the supply chain. Practices such as vendor-managed inventory (VMI); electronic point of sale (EPOS); collaborative planning, forecasting and replenishment (CPFR); etc. can surely have a positive effect on the application of green practices and become viable only with the use of ICT tools. Another critical aspect of information and communication systems, especially the internet, is the system-wide appreciation of environmental impacts of GSCM to various ecological systems and the natural environment and to bridge information gaps across the supply chain.
- Finally, GrICT systems support the decision-making process on the three business levels (operational, tactical and strategic).[89–91] The operational level refers to short-term decisions with horizons of hours or days, and which regard the daily operations of the enterprise. The tactical level refers to medium-term decisions with horizons expanding to weeks or months, and which regard the allocation/distribution and control of the enterprise's resources for the attainment of objective goals. The strategic level refers to long-term decisions whose horizons is more than one year, and which pose the long-term objectives of the enterprise. These decisions determine and establish the strategic plan which the enterprise will follow and designate the framework that operational

and tactical decisions will rest upon. Decisions across all levels relate and concern the:

- Selection of facilities (such as warehouses, distribution centers, etc.), their location, size and number. These decisions affect not only the total operating (logistics) costs, but also the energy use of facilities.
- Procurement planning including the selection of suppliers that have adopted green policies and practices or are located near the consumers (decreasing the corresponding distances).
- Distribution planning and transportation (selection of transportation mode, type and size of transportation/logistics unit, fuel, loading and routing, etc.), which affect both the environmental and economic performance because they have influence on the lead time (time needed to reach the final customer).
- Production management which also affects eco-efficiency because it determines the amounts of (raw) materials, energy and water needed, as well as the total amount of waste consumed/produced.[92]
- Inventory management and materials handling, which affects storage costs, especially in case of perishable (such as frozen, cooled or heated) products that require additional energy consumption.

These factors are illustrated in Figure 10.3.

Based on Figure 10.3, the various GrICT solutions can be categorized in terms of the support of the decision-making process in one or more business

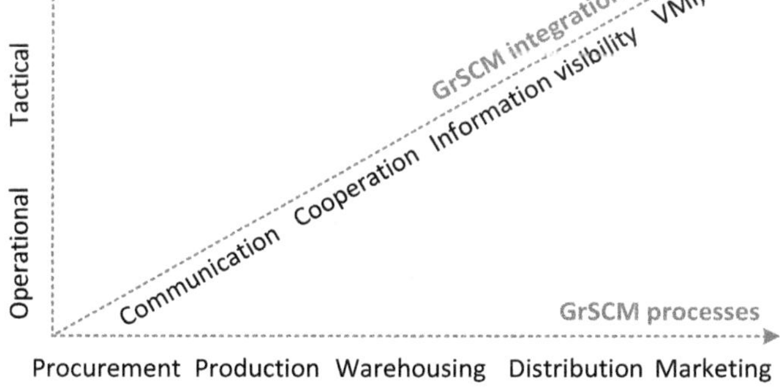

Figure 10.3 Affected areas of GrICT in GSCM

levels; the computerized automation of part or all of the operations or green initiatives, thus attaining technological and operational integration within the enterprise; and a level of integration with the other members/companies in the supply chain. Consequently, two main types of GrICT can be distinguished, the integrated solutions the development of which occurs in the context of a total (systems) approach to the green needs of a business and the integrated solutions aimed at the automation and optimization of individual operating areas, and/or the support of a particular type of decisions and for specific businesses/members of the green supply chain. Solutions coined as "best-of-breed" applications are usually the best market choices in their category made available by many and different manufacturers.

GrICT solutions

Figure 10.4 presents GrICT solutions that are available in the market.

An environmental management information system (EMIS) manages the daily functions performed by an enterprise that has opted to implement environmentally friendly techniques and practices. These informational systems have been developed to support the management of environmental management systems (EMS) that are based on international ISO standards. EMIS systems record and maintain real-time data that ensue from EMS activities, such as recording and monitoring of energy and water consumption of emissions and wastes, as well as of the activities relating to green practices. Moreover, they maintain a data repository with best practices, policies, standards, laws and regulations. EMIS systems support the set-up of special analyses, such as environmental cost/risk/impact assessment analysis and life cycle assessment (LCA), which is discussed in the following section. Additionally, they may be connected with other enterprise informational systems, such as enterprise resource planning, customer relationship management, supplier relationship management systems or quality and health and safety systems, as well as external information systems that maintain environmental data.

Life cycle assessment (LCA) tools are a technique used to quantify the environmental impact of a product or service in every stage of its life cycle (Figure 10.5).

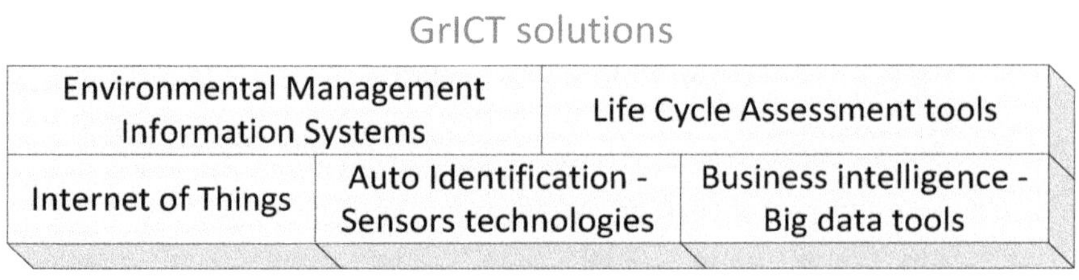

Figure 10.4 GrICT solutions in the market

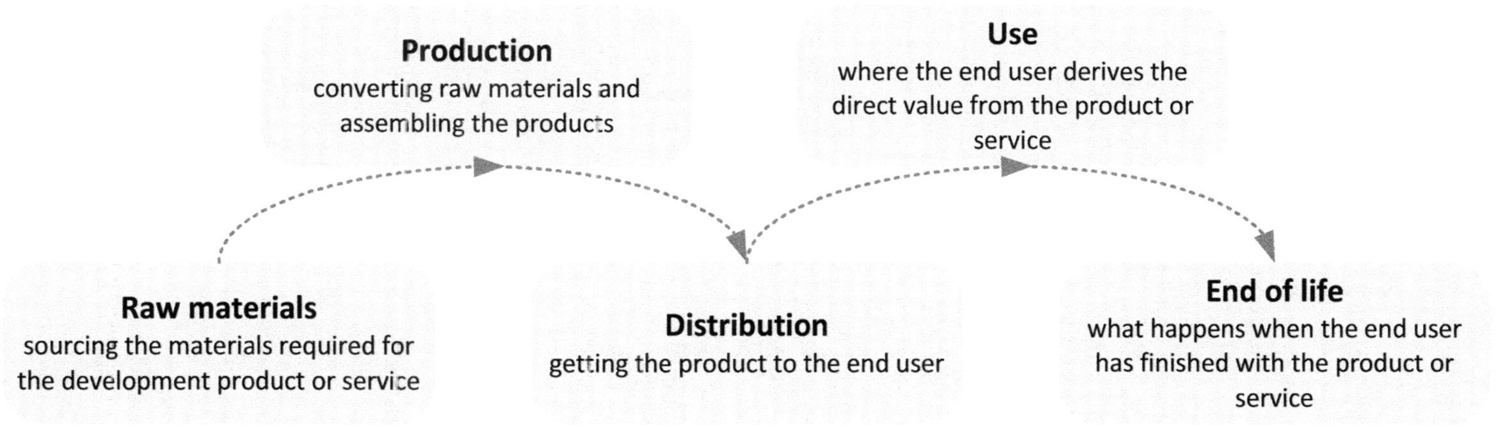

Figure 10.5 Stages in the life cycle of products and services, from raw material acquisition through end-of-life disposition (cradle-to-grave)

For example, the subject of each stage in the life cycle of petroleum products includes:

- ***Raw materials:*** This includes the extraction of raw materials from the earth, such as timber production or crude oil drilling. Also included in this stage is the transportation of raw materials from the point where these are extracted to the point where they will be processed.
- ***Production:*** Divided into two individual stages, the manufacture of materials and the manufacture of products. The manufacture of materials stage includes the conversion of a raw material to a form that can be used to make a final, complete product. For example, in order to produce polyethylene resin from crude oil, various manufacturing processes are required. Crude oil must be refined, ethylene must be produced from olefin by means of substitution reactions and then it must be polymerized in order to produce polyethylene. The transportation to the place where the product will be manufactured, as well as that between the individual manufacturing processes, are considered part of this stage, as are the packaging and standardization of the product.
- ***Distribution:*** Initially, this stage includes of all manufacturing processes needed for the packaging of the final product. It also includes all processes required for the packaging and distribution of the final product. Energy and waste ensuing from the transportation of the product to the retail points of sale or the consumer are also taken into account and measured at this stage of the product's life cycle.
- ***Use:*** This is the stage consumers are most familiar with. This stage includes energy and environmental waste requirements, which relate to the storage, use and maintenance/preservation of the product.
- ***End-of-life:*** This stage includes all energy and environmental waste requirements related to the disposal of the product as well as methods for managing waste after it is removed from the consumer, such as recycling, sanitary landfill and burning.

The LCA structure is defined and described by the International Organization for Standardization (ISO) in its 14000 series and pursuant to what is cited there, LC consists of the following four distinct phases:

- *Phase 1:* Goal definition and scoping to identify the LCA's purpose and the products of the study, and determine the boundaries.
- *Phase 2:* Life cycle inventory to quantify the energy and raw material inputs and environmental releases associated with each life cycle phase.
- *Phase 3:* Impacts analysis on human health and the environment.
- *Phase 4:* Reporting results to evaluate opportunities to reduce energy or material inputs, or environmental impacts, at each stage of the product life-cycle.

Although each phase serves an important function for the provision and collection of information, it must be clarified that it is not necessary for the analytical inventory and the assessment of the impact to be completed in order for the potential of environmental improvements to be identified. For example, the analytical inventory may be utilized by itself in order to identify potential for energy and raw materials savings, as well as for the reduction of waste.

During the performance of the preceding life cycle phase evaluations, the corresponding input and output data are collected using state-of-the-art LCA tools. The outputs of these tools provide very useful information for the support of green initiatives. An analytical data inventory can provide quantitative lists of energy and raw materials requirements, and atmospheric emissions, water waste and solid waste for a particular product, process, packaging, material or activity. Provided an analytical inventory has been performed and is deemed to be accurate with respect to its already designated subject and the limits of the system, its results can be directly utilized in order to identify areas of greater or lesser environmental loads, to support an impending impact assessment and to be part of a preliminary assessment of the impact. Following this, an impact assessment can be performed to quantify the effects on human health, as well as the effects on the health of the ecosystem that relate to specific harmful substances that have been identified at the stage of data inventorying. Critical factors for success include, on the one hand, the quality of data and, on the other, the integration of advanced algorithms and practices, but there are more critical points for the effective and efficient application of LCA tools. Moreover, a simple tool will be presented at the end of this chapter, in order for the reader to understand its use.

Initially, a **standardized exit unit must be designated** for every product, material and equipment participating in the production and management of a final product. All data can be presented relative to the production of a specific number of grams, kilograms or tons of said product. It is not necessary for the units used in its individual production stages to fit those of the final product.

The user of the LCA tools must then **establish energy and material requirements, as well as the environmental emissions** that contribute or are part of the manufacture of each product. This can be done with the help of a flowchart that will numerically determine the relations between independent production systems for the manufacture of the final product. The result is the calculation of energy and natural resources use, and of the environmental emissions of the entire system.

The next step is to **select the appropriate methodology**. The international literature on the subject includes of many methodologies that calculate environmental impact by taking into account the following:

- The entire production system.
- Relative contributions of the stages to the entire system.

- Relative contributions of the product's ingredients to the entire system.
- Data categories inside and between the stages, e.g. use of raw materials, energy consumption and environmental emissions.
- Groups of parametric data in a category, e.g. types of gaseous emissions, liquid and solid waste.
- Parametric data for a group, e.g. sulfur oxides, carbon dioxide, chlorine, etc.
- Geographic separation, if relevant to the survey, e.g. national against international.
- Changes over time.

The last step is to **publish the results**. There are two principal ways to present the results, tables and graphs.

Before their use is presented, one must note at this point that it will be purposeful to report the results with respect to the overall energy and in parallel to separate energy contributions for processing and for acquiring the raw materials. Solid waste can be distinguished between solid waste from use/consumption and industrial solid waste. Atmospheric and water hazardous substances must be reported separately. Atmospheric emissions, waste in the water and industrial solid waste can also be classified into processing-related emissions and those relating to fuel. Such classified reporting can contribute to the identification and, consequently, monitoring of energy consumption and environmental emissions.

The results of this detailed data-keeping can be presented in table form, rendering them easy to understand. This table can assume many forms, depending on the goal and subject of the survey. If the detailed record keeping performed was indented to assist in choosing the type of packaging to be used for a particular project, then the display of the results for the entire system will be the most appropriate way to present the results.

On the other hand, when an analysis is performed to determine the possibility of changing a packaging in order to reduce its emissions to the environment, it is important not only to present the overall results, but also the individual contributions of the parts of the packaging system.

The graphical representation of the data helps to reinforce tabulated data and facilitate their interpretation. Both bar as well as pie charts are valuable in assisting the reader to form a clear view and assimilate the information. For internal industrial use by product manufacturers, pie charts presenting a distinction between raw materials, processing and final use/disposal have proven useful in determining the potential for waste reduction. For external surveys, data must be presented in a way that will satisfy a fundamental criterion: clarity. To ensure clarity requires that the use/decision-maker formulates and answers queries relating to what kind of information and data needs to be conveyed by each graph. It may be necessary that a greater number of charts be presented with less data on each one. Following a simple consideration of the results, every reader must be in a position to comprehend the information the analyst intended to convey.

Internet of Things (IoT) technology allows to interconnect physical objects – called "things" or objects – in order to exchange information and functions between them. The types of "things" vary, depending on the applications and/or business sectors involved. For example:

- Sensors can be used to monitor moisture, temperature, etc. in a building or container.
- Sensors can be used to monitor the inventory level of products (physical quantities).
- Sensing nodes can also collect and monitor energy consumption, the quantity and flow of water, etc.
- Radar vision and RFID readers can sense the presence of an object, a person, etc.
- Locks/doors with open/closed circuits can trigger a building or container intrusion.
- Devices with embedded processing nodes can be used to control the status of vehicles (e.g. forklifts) or any kind of equipment.

Each "thing" bears an individual identity, while the interconnection and integration among them can be performed via remote commands – control topology and specialized software. The main idea is to "grab" the required data from the manufacturing and logistics processes.

In the case of green supply chain, IoT technology utilizes embedded sensors and communication technologies for the processes of manufacturing, storage, transportation and distribution, as well as conversely allowing data collection and coordination between "things".

The combination of IoT and cloud computing is also of great interest, as it allows the outsourcing of data processing and IT infrastructure, thus facilitating the ability to communicate with distributed systems and sensors in order to expand functionality and the allocation of resources.

Auto identifications/Sensors Technologies allow the automatic capture and storage of information at the time such information is created. These technologies include barcodes, radio frequency identification (RFID), smart cards, voice and vision identification, biometrics, etc. They can work together with sensors, which are usually small, portable battery-powered devices equipped with a microprocessor and memory for data storage. In the food industry, for example, RFID tags and sensors can be applied for the monitoring of fresh fruits and vegetables. They can be affixed to the inside of a container, monitor key variables such as temperature, humidity, etc., and submit them in other devises or information systems.

RFID technology is one of the automatic identification and data capture technologies that allow the automated capture and storage of information at the moment such information is created. These technologies also include other widely used technologies, such as barcodes, smart cards, voice and vision identification, biometrics, etc. RFID technology employs radio frequencies to read information stored in a small circuit known as RFID tag.[93]

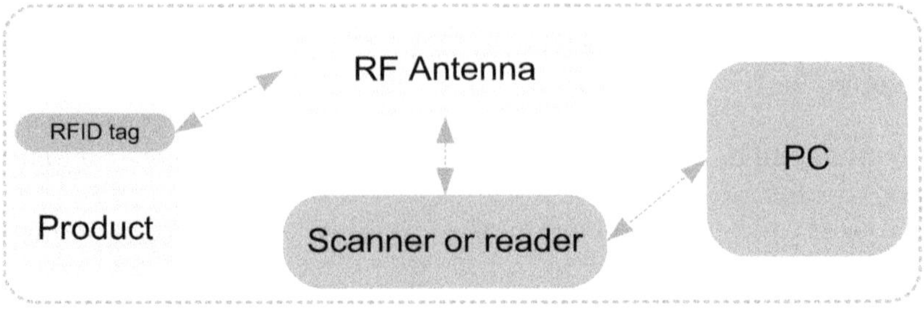

Figure 10.6 RFID functional elements

Tags can assume many different forms – they may be a small label on a product, such as a security film on an entry access card. They usually contain a microchip wherein data is stored and an antenna that allows them to receive and respond to the signal by a RFID transceiver. Specifically, there are three core parts to an RFID system: (a) an antenna; (b) the tags; and (c) a reader. These are illustrated in Figure 10.6.

The antenna is the device via which data is collected/transmitted to and from the tags. The tags are integrated circuits that can store data ranging from a few bits to several kilobytes. Tags can either be "read-only", namely programmed at their production stage without offering the capacity for further intervention, or "read-and-write", where their contents can be amended at any time. They can also be either "passive", which are activated when they are in the radio frequency field transmitted by a "reader", or "active", whereby they incorporate a transponder and a battery, and they transmit the information contained.

The reader is a device controlled by a computer and transmitting (via the antenna) RF waves to the tags, which must be located within a prescribed radius in order to be detected and activated. The same device also receives (again, via the antenna) and analyzes the information contained in each tag, which it then sends to a computer for further processing and storage.

RFID tags store information relating to the individuals or objects that bear such tags. Thus, they can be applied in a multitude of sectors where object recognition is necessary. They can, for example, be used in product packaging, in libraries, credit cards or even on a badge or identification document, such as government-issued identification cards, passports or drivers' licenses. One of their most common applications is the handling and transportation of products, whereby they can identify products either in transport or inside industrial units, whether such products are in pallets, warehouses or store shelves, thus replacing the barcode technology.

Among the advantages offered by RFID technology, we find the following: It requires no direct line of sight with the reader is necessary; it offers the capacity to store a large volume of data, to simultaneously read multiple tags; it offers accurate informing in real time; it facilitates the automated transmission of data to an informational system, the codification of a large volume of information, the ability to modify or amend the stored information and the reduction of manual labor; and it operates under adverse environmental conditions (barcodes, for example, are consumable and sensitive; they are, in other words, subject to easy erosion and destruction). RFID can help reduce thefts or losses, monitor the real life cycle of the products and check product quality. There are also some disadvantages that stand as the reasons for their non-adoption. These can be classified as technical, since the RFID systems by various manufacturers are not compatible with one another; financial, relating to the installation and use costs; and social, as RFID systems can infringe upon sensitive personal information.

Recent years have witnessed the emergence of smart containers that use electronic devices (such as RFID and global positioning systems, or GPS) and sensors to track and report data such as contents, unauthorized access and physical location in order to improve supply chain efficiencies and strengthen security. Figure 10.7 presents the box that consists of various sensors.

This system provides data to an application that enables the immediate and automatic identification and tracking of a container (Figure 10.8), in order to support: (a) identification of the exact position of the container (depiction on a map) from its entrance at the port to

Figure 10.7 Sensors inside the container

its placement in the ship; (b) recording of the control outcomes of the container content (in predefined or not controlled points); (c) recording of the status of the container in terms of temperature, humidity and violations; (d) notification of the decision-makers (operations and security officers) for an exceptional event (violation, move, not being checked, high temperature, etc.); and (e) exporting of reports and statistics regarding these factors.

Such a platform will today also include risk prevention systems to cover possible environmental accidents and emergencies improving drastically the efficiency of green practices and reducing waste for producers and distributors.

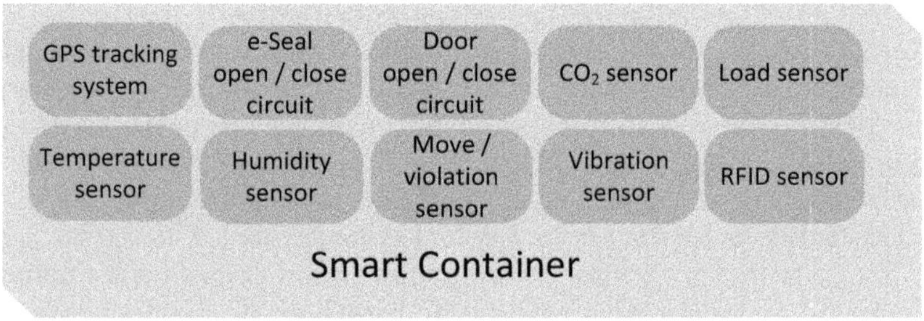

Figure 10.8 Sensors/embedded systems in "smart" container

Business intelligence

The goal of all ICT solutions is creation of a unified supply chain that allows for keeping suitable and reliable information with respect to products, materials and the energy consumed in order to support the decision-making process regarding the proper, efficient and effective management of GSCM operations.

In the modern business environment, business intelligence applications have to face significant challenges and satisfy specific requirements:

Big Data: There such a big volume of data that it cannot be managed using simple and traditional techniques. Aside from the usual sources of information, such as business information systems, there are also external sources of information originating from the systems of other members of the supply chain or from cloud computer infrastructures, as well as large amounts of data generated by sensors and embedded systems. Big data requires advanced data models and algorithms with prediction and adaptation capabilities.

***Business data is stored everywhere*:** Like on multiple systems (intra-business, subsystems, associates' systems, internet sites, social networks, etc.), and also in multiple forms (spreadsheets, non-structured documents, e-mail messages and other forms, e.g. XML markup language, etc.).

***The important role of business data semantics*:** In practice, this means that there is need for an agreement so as to render a common meaning for all business data (namely, shared definitions and unified semantics) by all collaborating organizations or businesses or parts thereof.

***Provision of fully configurable reports for areas of interest*:** Specifically, the creation of reports and evaluations based on indexes in business areas, such as financial analysis, using indexes and data relating to the basic financial statements, and HR analysis, using data and ratios relating to productivity, the development potential for human resources, etc. This also involves analysis of support functions for supplies using the relevant performance indexes and data in order to assess and evaluate their effectiveness.

A reliable, effective and efficient management of the available resources can improve the sustainability of the supply chain. Technologies such as data warehousing (DW), online analytical processing (OLAP), machine learning and data mining allow the decision-makers to have information in a more summarized manner and apply intelligence for optimized decisions.

The concept of DW is to simply extract operational transaction data into a separate repository and reorganize data in a de-normalized and aggregated fashion that is more appropriate for analysis and performing queries. OLAP provides the tools to perform interactive analysis against the warehoused data instead of operational files and databases. DW offers the benefit of integrated source data, which provides better data quality and consistency. It also enables data to be summarized and kept for historical reporting and analysis. The DW approach also reduces the decision support load on the operational transaction systems. The advantage of DW is that users can query data from across the enterprise. Using OLAP techniques, analysts create complex, multidimensional analyses and deliver to business users meaningful insights that might not be readily apparent (especially when applied with data mining).

Data mining is the process of searching for "hidden" relations and standards in the data kept in large data warehouses. These relations and standards depict valuable knowledge about the various entities of the database, such as customers, products, suppliers, employees, etc. Examples of such relations can be customers' reactions and behaviors with respect to particular offers, the compliance by suppliers to company demands and specifications, etc. Specific methodologies and tools are employed to implement this process, while its ultimate goal is to create models that will represent and will be in a position to anticipate or predict reality in the "best possible" way.

These software applications, technologies and analytical methodologies, which perform data analysis and support the decision-making process, constitute the business intelligence (BI) domain. BI captures organizational data from disparate sources and presents it to decision-makers in a user-friendly manner. It can be considered a value chain that converts primary data into information and useful knowledge, which leads to decisions and action. It exploits all of the software applications, practices, technologies and analytical methodologies that perform data analysis and provides real-time visibility and access to the right information, for the right individual at the right moment and in the right form, wherever their location may be, for each participant, in order to support enterprise competitiveness.

All of this will cater to and facilitate: (a) easy access and drawing of primary data from individual subsystems, but also external data kept in the systems of business associates (for example, the businesses or members of the supply chain); (b) homogenization and transformation of primary data to utilizable information by means of the application of common business rules; and (c) organization of information in data structures which are suitable for multidimensional analysis of large volumes of historic data.

The list of GrICT is not complete by citing the aforementioned technologies. New technologies are available in the market today: these include packaging technologies, in the context of the efforts to reduce the total amount of packaging materials, the amount of packaging waste while using environmentally friendly materials, etc.; transportation and distribution technologies using fossil fuel alternatives and biofuels such as ethanol and biodiesel in flexible fuel vehicles instead of gasoline and petroleum diesel, etc.

Making ICT solutions using green practices

The second axis relates to the correlation between green information and communication technologies (GrICT) and green supply chain management practices (GSCM) and regards the development of ICT solutions using green practices. Significant problems have arisen, according to Gartner Inc.; every year, the ICT industry generates 2% of global carbon dioxide (CO_2) emissions, while at the same time a typical personal computer (PC) consumes approximately 100–200 watts of electricity.[94] And since four billion PCs will be used by 2020, computer and ICT manufacturing in general have been rendered an industrial sector with significant impact on the natural environment.

In recent years, the continuously increasing interest of ICT consumers for green initiatives has pushed them to use their PCs and mobile devices accordingly and, in general, to behave in a way that they feel contributes to environmental protection. Moreover, it has pushed the companies of the ICT industry to follow green practices from the design and manufacturing stage of their products and for their management until their end-of-life.

The idea is simple and relates to the manufacture of computers and devices that are:

- Built from eco-friendly materials.
- Consume low power.
- Have computer power management (CPM) capabilities. This means devices that can power down when not in use or can help reduce the energy-related operating expenses of IT and reduce carbon emissions related to the global climate change.
- Are packaged using recyclable and environmentally sustainable materials.
- Can be recycled and reused after the end of its life cycle.

The last feature in particular relates to an existing and serious problem, the management of electronic waste (in the United States alone, roughly 100,000 devices are recycled on a daily basis, while the practice of burning the e-wastes causes the emission of harmful toxic substances), while at the same time being a great opportunity for the reuse of parts and for the recycling of raw materials such as gold, copper, silver, etc.[95]

ecoCycle: an easy life cycle analysis tool

This section is dedicated to the presentation of a simple, easy to use and available online LCA tool available at: www.greenagrochain.teithe.gr/ecoindicator/login.php. To use the application, users need to register. The administrator of the tool can grant access to users and statistically monitor the use of the tool. Below you can see a screenshot of the home page (Figure 10.9):

Data entry forms

To enter the data, the user is called to complete three consecutive steps relating to the production process, energy consumption and the management of the ensuing waste.

Production process data entry

In the first step (Figure 10.10), the type and quantity of raw materials used are identified, as well as the procedures required for the production process. Raw materials regard both the product and its packaging. The user of the tool will select the operating unit of the product, pursuant to which he/she wishes to perform the analysis. Additionally, transportation activities required to move the raw materials to and from the production unit are also stated at this step. The user will select the means of transportation employed

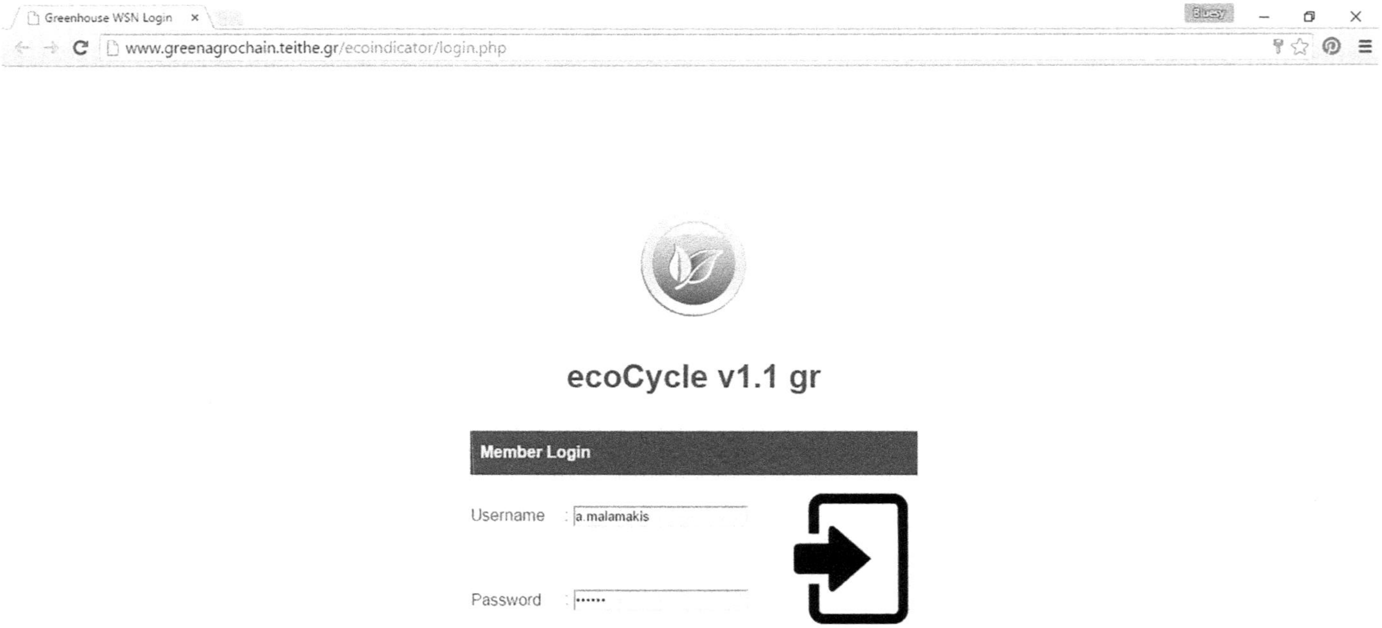

Figure 10.9 ecoCycle starting page

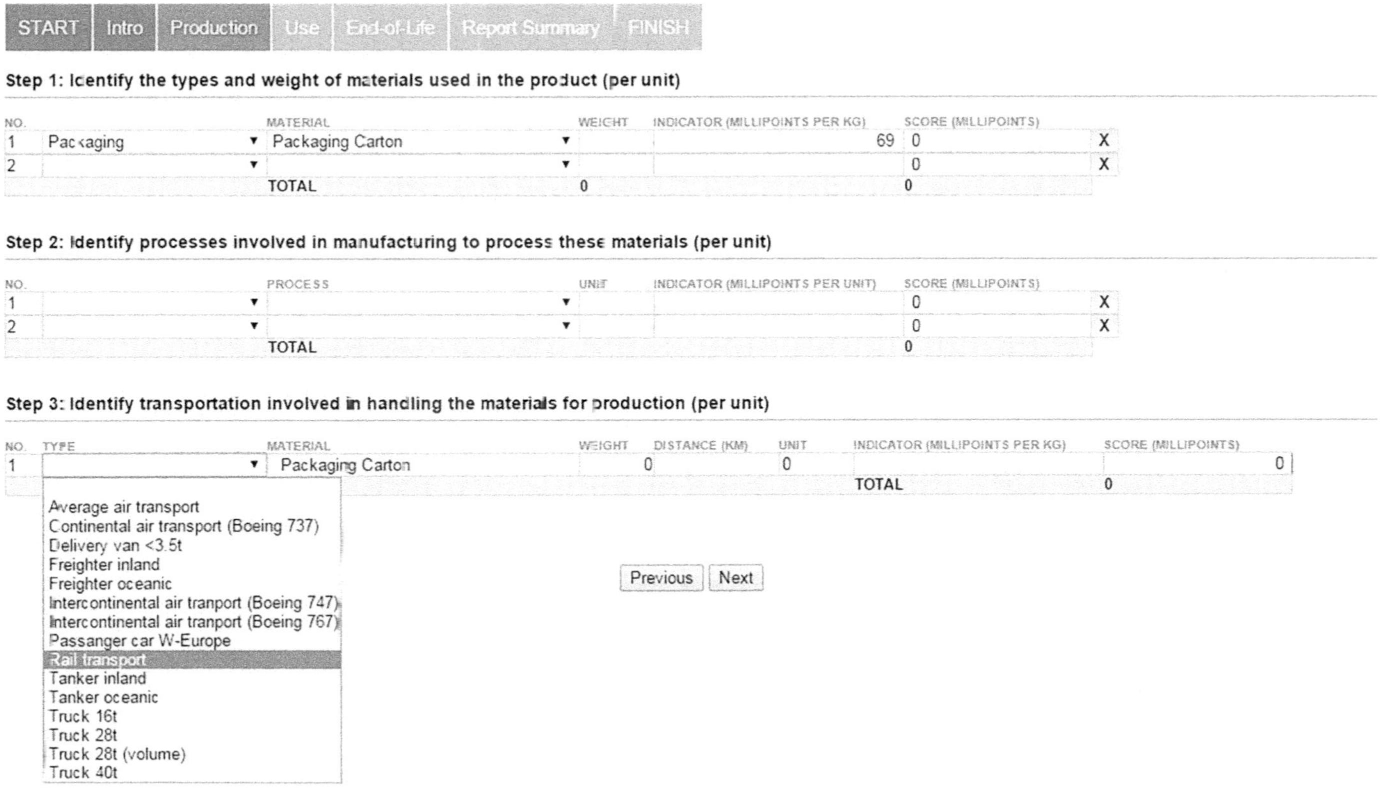

Figure 10.10 Entry of data relating to the production process

from a drop-down list, as well as the distance from the production unit. The selection of the materials is performed using the materials that were stated to be employed in the production process.

The user can revisit each stage to complete/correct the entry data. The tool will calculate the environmental impact for every entry data.

Energy consumption entry data

In the second entry stage, the user of the tool can enter data relating to energy consumption during the use/consumption of the product (Figure 10.11). The user defines the means of transportation for the product to the end user, as well as the consumption of energy (for example, consumption of electricity for freezing fruits until their final consumption) across the entire life cycle of the product.

The tool for the assessment of environmental impact has an incorporated database with different energy consumptions for all nations in the European Union. This way, the user can very accurately find the consumption data that relate to the product under assessment.

Waste management entry data

In the final step of data entry to the tool for the assessment of environmental impact (Figure 10.12), the user is called to quantify waste flow and the processes for the management of the waste that ensues at the end of the product's life cycle. At this stage, the user selects the method for the management of products after the end of their use. Options are available with respect to the final disposal of waste, as well as for the recycling of packaging materials. This tool offers users with the capacity to enter data relating to the transportation of waste to the locations for its management. This facilitates the ability to set limits for the system being examined, based on the needs of the analysis and the available data.

This tool has been created with an appropriate architecture which renders it easy to expand or/and specialize for specific materials and set individual goals. The emission and conversion factor libraries offer the potential to be expanded pursuant to the administrator's needs easily and expediently. The fields requesting information can be expanded during the use of the tool if a sufficient number of users finds there to be a need for the further analysis of individual categories.

Display of the tool's results

After the successful completion of all required data fields, the tool will display the results of the analysis (Figure 10.13). These results are presented both as absolute numbers as well as in the form of diagrams and pie charts. The analysis of the results is intended to offer users with the capacity to

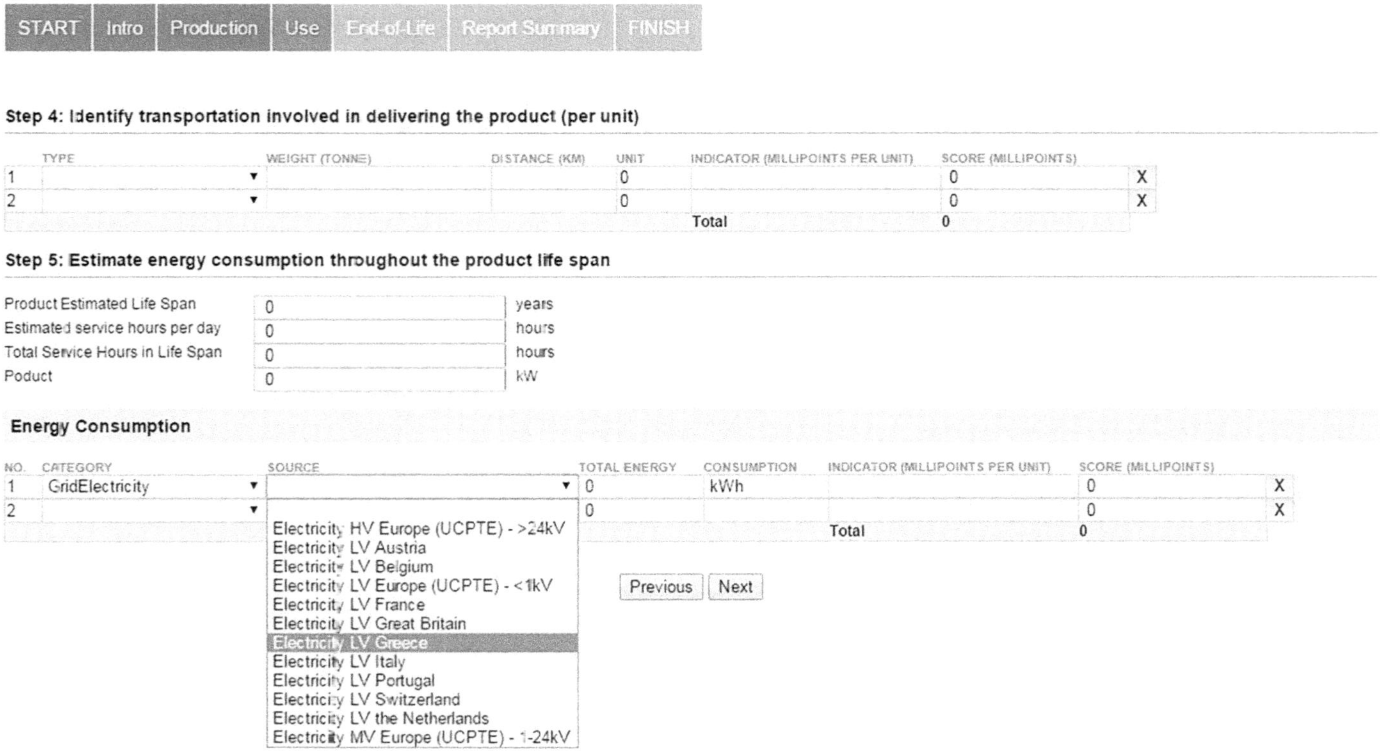

Figure 10.11 Data entry relating to energy consumption

Figure 10.12 Data entry relating to waste management

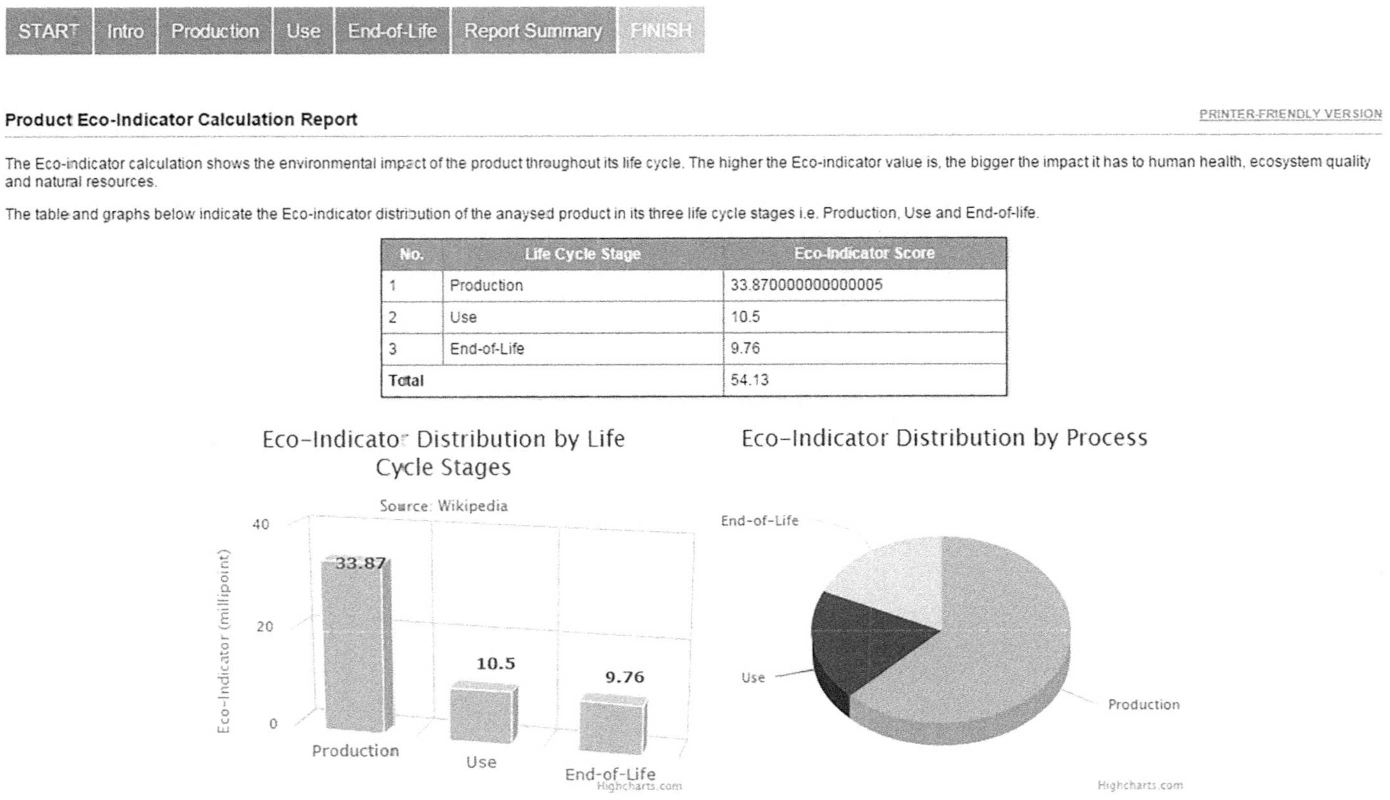
START | Intro | Production | Use | End-of-Life | Report Summary | FINISH

Product Eco-Indicator Calculation Report

PRINTER-FRIENDLY VERSION

The Eco-indicator calculation shows the environmental impact of the product throughout its life cycle. The higher the Eco-indicator value is, the bigger the impact it has to human health, ecosystem quality and natural resources.

The table and graphs below indicate the Eco-indicator distribution of the anaysed product in its three life cycle stages i.e. Production, Use and End-of-life.

No.	Life Cycle Stage	Eco-Indicator Score
1	Production	33.870000000000005
2	Use	10.5
3	End-of-Life	9.76
Total		54.13

Figure 10.13 Results of calculations

evaluate which point in the product's life cycle is the one contributing the most to its environmental impact. This way, the user can then direct environmental strategy towards the rationalization of the product's life cycle.

Chapter summary

Technological advances provide valuable tools to the green supply chain framework. Green information and communication technologies emerged as a sector that offers a versatile – and at the same time, stable – base upon which green practices can be built and perform optimally. The integration of GrICT revealed areas of interest and brought forth the development of novel offerings to the complex supply chain ecosystem. GrICT integrated with GSCM enhances established tools (e.g. life cycle analysis) and introduces innovative concepts such as environmental management information systems and integrated elements in other information systems. In this chapter, we focused on the definition of GrICT and the available smart ICTs and sustainable ICTs. Moreover, we examined the role of GrICT and how they are related to green supply chain management, and analyzed the areas most affected by GrICT in the supply chain of an enterprise.

References

1 Fang, C. & Zhang, J. Performance of green supply chain management: A systematic review and meta analysis. *J. Clean. Prod.* **183**, 1064–1081 (2018).

2 Engels, D. *Alexander the Great and the logistics of the Macedonian army*, 1st edition. (University of California Press, 1978).

3 Lyon, T. P. & Maxwell, J. W. Corporate social responsibility and the environment: A theoretical perspective. *Rev. Environ. Econ. Policy* **2**, 240–260 (2008).

4 Dauvergne, P. & Lister, J. Big brand sustainability: Governance prospects and environmental limits. *Glob. Environ. Chang.* **22**, 36–45 (2012).

5 Rothenberg, S., Pil, F. K. & Maxwell, J. Lean, green, and the quest for superior environmental performance. *Prod. Oper. Manag.* **10**, 228–243 (2001).

6 Mokhtarian, P. L. A conceptual analysis of the transportation impacts of B2C e-commerce. *Transportation (Amst).* **31**, 257–284 (2004).

7 Walker, H., Di Sisto, L. & McBain, D. Drivers and barriers to environmental supply chain management practices: Lessons from the public and private sectors. *J. Purch. Supply Manag.* **14**, 69–85 (2008).

8 Clinton, W. J. & Gore A. *Reinventing environmental regulation.* (National Service Center for Environmental Publications, 1955).

9 Chin, T. A., Tat, H. H. & Sulaiman, Z. Green supply chain management, environmental collaboration and sustainability performance. *Procedia CIRP* **26**, 695–699 (2015).

10 Peattie, K. & Charter, M. Green marketing. In *The marketing book* 726–755. (Butterworth-Heinemann, 2003).

11 Blome, C., Hollos, D. & Paulraj, A. Green procurement and green supplier development : Antecedents and effects on supplier performance. *Int. J. Prod. Res.* **52**, 32–49 (2014).

12 Novack, R. A. & Simco, S. W. The industrial procurement process: A supply chain perspective. *J. Bus. Logist.* **12**, 145–167 (1991).

13 ECO-Buy Limited. *Sustainable procurement guide.* (Australian Government, 2013). doi:10.3403/9780580698613

14 OECD. *Going green: Best practices for sustainable procurement.* 2015. Available at: https://www.oecd.org/gov/ethics/Going_Green_Best_Practices_for_Sustainable_Procurement.pdf

15 European Commission. *Buying green! A handbook on green public procurement.* (Publications Office of the European Union, 2016). doi:10.2779/246106

16 Chiu, (Anthony) Shun Fung, Tan, R. R. & Siriban-Manalang, A. B. Sustainable consumption and production for Asia: Sustainability through green design and practice. *J. Clean. Prod.* **40,** 1–5 (2013).
17 King, A. A. & Lenox, M. J. Lean and green? An empirical examination of the relationship between lean production and environmental performance. *Prod. Oper. Manag.* **10,** 244–256 (2009).
18 Baines, T., Brown, S., Benedettini, O. & Ball, P. Examining green production and its role within the competitive strategy of manufacturers. *J. Ind. Eng. Manag.* **5,** 53–87 (2012).
19 Brugha, R. & Zsuzsa, V. Review article stakeholder analysis: A review. *Health Policy Plann* **15,** 239–246 (2000).
20 Bochtis, D. D., Sørensen, C. G. C. & Busato, P. Advances in agricultural machinery management: A review. *Biost. Eng.* **126,** 69–81 (2014).
21 Rodias, E. *et al.* Energy savings from optimised in-field route planning for agricultural machinery. *Sustainability* **9,** 1956 (2017).
22 Zangheri, P., Economidou, M. & Kona, A. *Summary report: Analysis of the annual reports 2017 under the Energy Efficiency Directive* (European Commission, 2017). Available at: http://publications.jrc.ec.europa.eu/repository/bitstream/JRC108810/eed_annual_reports_2017_final.pdf
23 German, J. *Hybrid vehicles technology development and cost reduction* (International Council of Clean Transportation, 2015). Available at: https://www.theicct.org/sites/default/files/publications/ICCT_TechBriefNo1_Hybrids_July2015.pdf
24 De Santi, G., Edwards, R., Szekeres, S. & Mahieu, V. *Biofuels in the European context: Facts and uncertainties* (European Commission, 2008). Available at: https://ec.europa.eu/jrc/sites/jrcsh/files/jrc_biofuels_report.pdf
25 Nigel, H. *McDonald's to recycle cooking oil for fuel. 2007.* Available at: https://uk.reuters.com/article/uk-mcdonalds-biodiesel/mcdonalds-to-recycle-cooking-oil-for-fuel-idUKMOL23573620070702 (2012)
26 McDonald's. *McDonald's corporate responsibility and sustainability report.* Available at: https://mcdonalds.com.au/sites/mcdonalds.com.au/files/MCD_CRS_Complete.pdf
27 Jensen, M. F., Bochtis, D. & Sørensen, C. G. Coverage planning for capacitated field operations, part II: Optimisation. *Biosyst. Eng.* **139,** 149–164 (2015).
28 Jensen, M. A. F., Bochtis, D., Sorensen, C. G., Blas, M. R. & Lykkegaard, K. L. In-field and inter-field path planning for agricultural transport units. *Comput. Ind. Eng.* **63,** 1054–1061 (2012).
29 Bochtis, D. D. & Sørensen, C. G. The vehicle routing problem in field logistics: Part II. *Biosyst. Eng.* **105,** 180–188 (2010).
30 Gesing, A. & Wolanski, R. Recycling light metals from end-of-life vehicle. *JOM* **53,** 21–23 (2001).
31 Burnham, A., Wang, M. Q. & Wu, Y. *Development and applications of GREET 2.7 – The transportation vehicle-cycle model.* (2006). doi:10.2172/898530
32 Dablanc, L. Goods transport in large European cities: difficult to organize, difficult to modernize. *Transp Res Part A Policy Pract.* **41** (3), 280–285 (2007).
33 European Commission. *White paper on transport roadmap to a single European transport area – Towards a competitive and resource-efficient transport system.* (2011). doi:10.2832/30955
34 Hellström, D. & Saghir, M. Packaging and logistics interactions in retail supply chains. *Packag. Technol. Sci.* **20,** 197–216 (2007).

35 Bjørner, T. B., Hansen, L. G. & Russell, C. S. Environmental labeling and consumers' choice – An empirical analysis of the effect of the Nordic Swan. *J. Environ. Econ. Manag.* **47,** 411–434 (2004).
36 Teisl, M. F., Roe, B. & Hicks, R. L. Can eco-labels tune a market? Evidence from dolphin-safe labeling. *J. Environ. Econ. Manag.* **43,** 339–359 (2002).
37 Rodriguez, A. *There are more than 450 meanings behind "green" labels.* 2015. Available at: https://qz.com/521251/there-are-more-than-450-meanings-behind-green-labels/
38 European Commission. *Energy efficient products.* n.d. Available at: https://ec.europa.eu/energy/en/topics/energy-efficiency/energy-efficient-products
39 *Environmental labels and declarations: How ISO standards help.* n.d. Available at: https://www.iso.org/files/live/sites/isoorg/files/archive/pdf/en/environmental-labelling.pdf
40 Baker, S. The evolution of European Union environmental policy. *Polit. Sustain. Dev.* 89 (2012).
41 Berruto, R., Busato, P., Bochtis, D. D. & Sørensen, C. G. Comparison of distribution systems for biogas plant residual. *Biomass Bioenergy* **52,** 139–150 (2013).
42 European Commission. Directive 2012/19/EU of the European Parliament and of the Council of 4 July 2012 on waste electrical and electronic equipment (WEEE). OJ L 197, 38–71 (2012).
43 Achillas, C. End of life management of industrial products based on environmental and economic criteria and optimisation of the reverse supply chain network. PhD Thesis, Mechanical Engineering Department, Aristotle University Thessaloniki, Greece, June (2009).
44 Janz, A. & Bilitewski, B. Auswirkung des ElektroG auf den schadstoffeintrag im restabfall. *Tech. Univerität Dresden, Brennpkt. ElektroG Umsetzung – Defizite – Notwendigkeiten, Dresden, Ger.* **23** (2009).
45 Chancerel, P. & Rotter, S. Recycling-oriented characterization of small waste electrical and electronic equipment. *Waste Manag.* **29,** 2336–2352 (2009).
46 Janz, A. & Bilitewski, B. *WEEE in and outside Europe-hazards, challenges and limits.* (na, 2009).
47 Innes, S. *Developing tools for designing out waste pre-Site and onsite. In Proceedings of Minimizing Construction Waste Conference: Developing Resource Efficiency and Waste Minimization in Design and Construction.* (New Civil Engineer, London, UK, 2004).
48 Aidonis, D. Applied operations research methodologies for the optimal design and operation of reuse and recycling networks of construction and demolition materials. PhD thesis, Mechanical Engineering Department, Aristotle University Thessaloniki, Greece, (June 2009).
49 Osmani, M., Glass, J. & Management, A. P.-W. undefined. Architects' perspectives on construction waste reduction by design. *Elsevier* (2008).
50 Cheung, C. M., Wong, K. W., Fan, C. N. & Poon, C. S. *Reduction of construction waste: Final report.* (Hong Kong Polytechnic, 1993)
51 Kourmpanis, B. *et al.* Preliminary study for the management of construction and demolition waste. *Waste Manag. Res.* **26,** 267–275 (2008).
52 Greer, D. Building the deconstruction industry, *BioCycle* **45** (11), 36 (2004).
53 Webster, R. & Napier, T. Deconstruction and reuse: Return to true resource conservation and sustainability. *Fed. Facil. Environ. J.* **14,** 127–143 (2003).
54 Brandon, P. S. & Lombardi, P. *Evaluating sustainable development in the built environment.* (Wiley, 2010).

55 Curwell, S. R., Deakin, M. & Symes, M. *Sustainable urban development: The environmental assessment methods*. **2**, (Taylor & Francis, 2005).
56 Deakin, M., Mitchell, G., Nijkamp, P. & Vreeker, R. *Sustainable urban development volume 2: The environmental assessment methods*. (Routledge, 2007).
57 Marvin, S. & Guy, S. Creating myths rather than sustainability: The transition fallacies of the new localism. *Local Environ.* **2**, 311–318 (1997).
58 van den Bergh, J. C. J. & Verbruggen, H. Spatial sustainability, trade and indicators: An evaluation of the 'ecological footprint'. *Ecol. Econ.* **29**, 61–72 (1999).
59 Lombardi, P. & Brandon, P. Towards a multimodal framework for evaluating the built environment quality in sustainability planning, in *Evaluation of the built environment for sustainability*, Eds. P. Brandon, V. Bentivegna & P. Lombardi. (Chapman & Hall, 1997).
60 Nijkamp, P. & Pepping, G. A meta-analytical evaluation of sustainable city initiatives. *Urban Stud.* **35**, 1481–1500 (1998).
61 *U.N. Report of the World Commission on Environment and Development: Our common future. Transmitted to the General Assembly as an Annex to document A/42/427 – Development and International Cooperation: Environment*. 1987. Available at: http://www.exteriores.gob.es/Portal/es/PoliticaExteriorCooperacion/Desarrollosostenible/Documents/Informe%20Brundtland%20(En%20ingl%C3%A9s).pdf
62 Hutchins, M. J. & Sutherland, J. W. An exploration of measures of social sustainability and their application to supply chain decisions. *J. Clean. Prod.* **16**, 1688–1698 (2008).
63 Poveda, C. A. & Lipsett, M. Journal of sustainable development JSD. *Journal of Sustainable Development* **4**, (CCSE, 2011).
64 Pearce, D. W., Atkinson, G., Mourato, S. & Organisation for Economic Co-Operation and Development. *Cost-benefit analysis and the environment : Recent developments*. (Organisation for Economic Co-operation and Development, 2006).
65 Cummings, R. G. & Taylor, L. O. Unbiased value estimates for environmental goods: A cheap talk design for the contingent valuation method. *Am. Econ. Rev.* **89**, 649–665 (1999).
66 Rosen, S. Hedonic prices and implicit markets: Product differentiation in pure competition. *J. Political Econ.* 82 (1), 34–55 (1974).
67 Munda, G. Multiple criteria decision analysis and sustainable development, in *Multiple criteria decision analysis: State of the art surveys*, Eds. M. Ehrgott & J. R. Figueira 953–986 (Springer, 2005).
68 Sinivuori, P. & Saari, A. MIPS analysis of natural resource consumption in two university buildings. *Build. Environ.* **41**, 657–668 (2006).
69 Saling, P. *et al.* Eco-efficiency analysis by basf: The method. *Int. J. Life Cycle Assess.* **7**, 203–218 (2002).
70 Tugnoli, A., Santarelli, F. & Cozzani, V. An approach to quantitative sustainability assessment in the early stages of process design. *Environ. Sci. Technol.* **42**, 4555–4562 (2008).
71 Rees, W. & Wackernaegel, M. Urban ecological footprints: Why cities cannot be sustainable and why they are a key to sustainability. *Environ. Impact Assess. Rev.* **16**, 223–248 (1996).
72 Traverso, M., Finkbeiner, M., Jørgensen, A. & Schneider, L. Life cycle sustainability dashboard. *J. Ind. Ecol.* **16**, 680–688 (2012).
73 Alberti, M. & Susskind, L. Managing urban sustainability: An introduction to the special issue. *Environ. Impact Assess. Rev.* **16**, 213–221 (1996).

74 Waas, T. *et al.* Sustainability assessment and indicators: Tools in a decision-making strategy for sustainable development. *Sustain.* **6,** 5512–5534 (2014).
75 Bochtis, D. D., Sørensen, C. G. & Green, O. A DSS for planning of soil-sensitive field operations. *Decis. Support Syst.* **53,** 66–75 (2012).
76 Dahl, A. L. Achievements and gaps in indicators for sustainability. *Ecol. Indic.* **17,** 14–19 (2012).
77 Moldan, B., Janoušková, S. & Hák, T. How to understand and measure environmental sustainability: Indicators and targets. *Ecol. Indic.* **17,** 4–13 (2012).
78 Hezri, A. A. Sustainability indicator system and policy processes in Malaysia: A framework for utilisation and learning. *J. Environ. Manag.* **73,** 357–371 (2004).
79 Meadows, D. *Indicators and information systems for sustainable development.* (The Sustainability Institute, 1998).
80 Bebbington, J., Brown, J. & Frame, B. Accounting technologies and sustainability assessment models. *Ecol. Econ.* **61,** 224–236 (2007).
81 http://sedac.ciesin.columbia.edu/data/collection/esi/
82 http://sedac.ciesin.columbia.edu/data/collection/epi
83 Sørensen, C. G. et al. Conceptual model of a future farm management information system. *Comput. Electron. Agric.* **72** (1), 37–47 (2010).
84 Sørensen, C. G., Pesonen, L., Bochtis, D. D., Vougioukas, S. G. & Suomi, P. Functional requirements for a future farm management information system. *Comput. Electron. Agric.* **76,** 266–276 (2011).
85 de Camargo Fiorini, P. & Jabbour, C. J. C. Information systems and sustainable supply chain management towards a more sustainable society: Where we are and where we are going. *Int. J. Inf. Manag.* **37,** 241–249 (2017).
86 Rodias, E., Berruto, R., Bochtis, D., Busato, P. & Sopegno, A. A computational tool for comparative energy cost analysis of multiple-crop production systems. *Energies* **10,** 831 (2017).
87 Sopegno, A. *et al.* Model for energy analysis of miscanthus production and transportation. *Energies* **9,** 392 (2016).
88 Busato, P., Sopegno, A., Berruto, R., Bochtis, D. & Calvo, A. A web-based tool for energy balance estimation in multiple-crops production systems. *Sustainability* **9,** 789 (2017).
89 Zhou, K., Jensen, A. L., Bochtis, D. D. & Sørensen, C. G. Quantifying the benefits of alternative fieldwork patterns in a potato cultivation system. *Comput. Electron. Agric.* **119,** 228–240 (2015).
90 Orfanou, A. *et al.* Scheduling for machinery fleets in biomass multiple-field operations. *Comput. Electron. Agric.* **94,** 12–19 (2013).
91 Pavlou, D. *et al.* Functional modeling for green biomass supply chains. *Comput. Electron. Agric.* **122,** 29–40 (2016).
92 Bochtis, D. D. *et al.* A flow-shop problem formulation of biomass handling operations scheduling. *Comput. Electron. Agric.* **91,** 49–56 (2013).
93 Finkenzeller, K. *Rfid handbook: Fundamentals and applications in contactless smart cards, radio frequency identification and near-field communication.* (Wiley, 2014).
94 Stamford, C. (2007) Gartner Estimates ICT Industry Accounts for 2 Percent of Global CO2, online, available at: https://www.gartner.com/newsroom/id/503867.
95 Electronics TakeBack Coalition (2014) Facts and Figures on E-Waste and Recycling, online, available at: http://www.electronicstakeback.com/wp-content/uploads/Facts_and_Figures_on_EWaste_and_Recycling.pdf.

Index

Note: Page numbers in *italic* indicate a figure and page numbers in **bold** indicate a table on the corresponding page.

agile supply chain 8
Alexander the Great 5–6
AMR Research and Forrester 63
analytic network process (ANP) 140
application steps of set pair analysis 55–60; assessing alternatives based on quantitative criteria 58–59; assessing alternatives by criterion A 57; assessing alternatives by criterion B **57**; assessing alternatives by criterion C **57**; assessing alternatives by criterion D **57**; assessing alternatives by criterion E **58**; calculation of criteria weightings **56**; evaluating the selected criteria 56; final assessment of alternatives **61**; final assessment of alternatives based on qualitative and quantitative criteria **59**; overall assessment of alternatives based on qualitative criteria 58; overall assessment of alternatives based on qualitative criteria **59**; rating of alternatives based on quantitative criteria **59**; scoring alternatives for each criterion 56–58; selection of assessment criteria 55; weighing qualitative criteria in pairs **56**
auto identifications/sensors technologies 155–158

Blue Angel 92
bounded rationality implications 19, 37
British Standards Institution 26
BS 7750 26
B2B *see* business-to-business
business intelligence 158–160; big data 158; business data semantics 159; business data stored everywhere 159; provision of fully configurable reports for areas of interest 159
business-to-business (B2B) 13, 36, 93

CBA *see* cost-benefit analysis
clean technologies 15, 17, 19, 37; access prohibiting factors 24–25
codes of practice 33, 37
community stakeholder power 35, 37
conceptualization stage 67, 70–72
construction and demolition waste (CDW) 99, 115, 119, *122*, 135
contingent valuation method (CVM) 139
conventional demolition 130, 131, 132, *133*
corporate social responsibility (CSR) 10, *10*, 11
cost-benefit analysis (CBA) 19, 138–139
counteracting policies and subsidies 25, 37

decision-making 136–45; analytic network process (ANP) 140; assessment indicators 143–144; assessment methods 136–137, *137*; contingent valuation method (CVM) 139; cost-benefit analysis (CBA) 19, 138–139; eco-efficiency analysis (ECOEFA) 141–142; eco-efficiency portfolios 142; environmental social

and economic impact analysis 138; frameworks and methods 138–143; Global Ecolabeling Networking (GEN) 50; hedonic pricing method (HPM) 139; life cycle assessment (LCA) 88, 141, 150–154, 161; life cycle sustainability dashboard (LCSD) 142–143; material intensity per service unit 140; multi-criteria analysis (MCA) 55, 111, 137, 139, 140; quantitative assessment of sustainability indices (QASI) 142; strategic environmental assessments 137, 138; sustainability reporting guidelines (SRG) 141; US National Environmental Policy Act 138
department-network stage (stakeholder analysis) 67, 68–69
door to door (DTD) 12

ECDW *see* excavation, construction and demolition waste management
ecoCycle 161–168, *162*; calculation results *167*; data entry forms 161–164; data entry relating to energy consumption *165*; data entry relating to waste management *166*; display of tool's results 164–168; entry of data relating to production process *163*
eco efficiency analysis (ECOEFA) 141
eco-efficiency portfolios 142
ecolabels 50, 94; environmental labeling applications *95*, 95–96; environmental labeling process *94–95*; Type I environmental labeling—ecolabel (ISO 14024); Type II environmental labeling—environmental claims (ISO 14021) 93; Type III environmental labeling—environmental product declaration (ISO 14021) 93–94; *see also* Blue Angel; Ecolabel Flower; EU Ecolabel Product Catalogue; Global Ecolabeling Networking
Ecolabel Flower 89
ecological footprint (EF) 140, 142
economic environment fluctuations 26, 37
economic incentives 32–33, 37
education and training 33–34, 37
EEE *see* electrical and electronic equipment
EIA *see* environmental impact analysis
electrical and electronic equipment (EEE) 99, 100, 102, 103, 109, 111, 112, 113, 114, 115; cutting/separation 106; disassembly 107–108; disposal 109; hazardous materials **103**; heat treatment 108; reuse 105, 106; stream *104*
EMS *see* environmental management systems
enablers and barriers 14–37; external barriers 22–26, *23*, 37 (*see also* external barriers); external motivators and drivers 30–37 (*see also* external motivators and drivers); internal barriers 15–22, *16*, 37 (*see also* internal barriers); internal motivators and drivers 9–10, 26–30, *27*, 37 (*see also* internal motivators and drivers)
energy intensity 71
Energy Label 89
Energy Star 47, 92, 93
enhancing efficiency 30, 37
environmental accounting 29–30, 37
environmental auditing 35–36, 37
environmental awareness 34, 65, 72, 85–86, 91, 118
environmental functionalities 72
environmental impact analysis (EIA) 138
environmental issues 8–9, 10, 14, 17, 22, 28, 34, 35, 36, 37, 75, 78, 88, 90
environmental leadership 28–29, 37
environmental management and auditing scheme (EMAS) 26, 48, 49
environmental management systems (EMS) 32, 34, 35, 37, 95, 150; continuous improvement 26–28
environmental product declarations 93–94
environmental reporting 29, 37
EU Ecolabel Product Catalogue 47
EU Energy Star 47, 92–93
European Single Procurement Document 51
European Union sustainability guidelines 84–85
evolution of traditional supply chains 1, 3, 6–8, *7*, 24

excavation, construction and demolition waste management (ECDW) 115–123, *116*, *117*, 127, 130, 135; alternative demolition procedures 131–132; construction phase 127–129; construction waste management options *123–124*; conventional demolition 130, 131, 132, *133*; decision-making methodological framework 123–132; demolition phase 129–131; design phase 124–127; dimensions of management 118–123; economic dimensions 118–119; environmental dimensions 119–120; legislative dimensions 123; material and information flow in the construction industry supply chain *124*; parameters that need to be considered in the design of a building *126*; ranking of management policies *120*; selective deconstruction 125, 127, 130, 131–132, *133*, *134*; technological dimensions 120–123, *121*
external barriers 22–26, *23*, 37; clean technologies access prohibiting factors 24–25; counteracting policies and subsidies 25; economic environment fluctuations 26, 37; lack of suitable markets for recycled goods 26; regulatory approach's fail points 22–24, 37; restricted access to external funding 25, 37
external funding 25, 37
external motivators and drivers 30–37, *31*; codes of practice 33, 37; community stakeholder power 35, 37; economic incentives 32–33, 37; education and training 33–34, 37; environmental auditing 35–36, 37; financial stakeholders 29, 35, 37; green consumers 11, 36, 37, 90, 112; international trade 7, 36, 37; peer networks 34, 35, 37; regulation fostering pollution prevention innovation 31–32, 37; self-regulation 32, 37; supply chain pressure 34–35, 37

financial accounting under- and misrepresentation of environmental benefits and costs 17–18
financial performance 11, 18, 32, 35, 37; pressure 19–20
financial stakeholders 29, 35, 37
fine-tune performance 72

GHG emissions *see* greenhouse gas (GHG) emissions
Global Ecolabeling Networking (GEN) 50
Global Reporting Initiative 141
green consumers 11, 36, 37, 90, 112
greenhouse gas (GHG) emissions 54, 63n1, 76, *76*, 80
green information and communication technologies (GrICT) 4, 146, 147, 148, 149, 160, 168; affected areas in GSCM *149*; aspect and GSCM relations *146*; aspects in GSCM *147*; solutions *150*, 150–158, 160–161
green packaging 87–96; benefits and disadvantages of environmental labels 90–91, *90*; ecolabels 91–94 (*see also* ecolabels); environmental labeling and labels 88–89
green practices in procurement *43(see also green procurement)*
green practices in transportation/distribution 77–84, *83*; adoption and implementation of telematics systems and technologies 82–83; adoption of eco-driving habits 83; appropriate packaging 83; better route design 80; designing green-sustainable distribution networks 81; design of efficient distribution networks 77–79; environmental indicators and benefits in context of local food supply chains **79–80**; European Union sustainability guidelines 84–85; increase of utilization degree 80–81; preference and use of non-polluting means of transport 77; recycling of old vehicles 80; use of an energy efficient transportation and distribution fleet 77
green procurement 38–63; barriers to broader adoption 60–62; benefits *43*; best practices for application in enterprises **42**; best practices for application in public sector **42**; definition 39–41; factors that contribute to increased interest

41–44; lifecycle 44–60 (*see also* life cycle of green procurement)
green production 64–73; conceptualization stage 70–72; department-network stage (stakeholder analysis) 68–69; design 65–66, *71*; eco-profile stage 68; life cycle assessment methods **69**; overview stage 67–68; quantification stage 69–70; stages 66–73, *67*, 72–73
green supply chain framework 1–4, *2*, 168
green supply chain management (GSCM) 1, 3, 4, 11, 14, 18, 19, 21, 22, 24, 26, 30, 32, 144, 146, *146*, 147, *147*, 148, *149*, 158, 160, 168
green technologies 4, 21, 22, 24, 33, 37
green transportation *see* green practices in transportation/distribution
GrICT *see* green information and communication technologies
GSCM *see* green supply chain management

harmful substances 71, 89, 109, 153
hedonic pricing method (HPM) 139
hybrid 8, *9*, 77, 94

ICTs *see* information and communication technologies
inbound logistics (green procurement) 38–63; barriers to broader adoption 60–62; benefits *43*; best practices for application in enterprises **42**; best practices for application in public sector **42**; definition 39–41; factors that contribute to increased interest 41–44; lifecycle 44–60 (*see also* life cycle of green procurement)
information and communication technologies (ICTs) 4, 146, 147; smart 146; solutions 147–160; solution using green practices 160–168; sustainable 146; *see also* green information and communication technologies (GrICT)
intelligent transportation systems 82
internal barriers 15–22, *16*, 37; bounded rationality implications 19, 37; challenges in transitioning to green technologies 21–22, 37; financial accounting under- and misrepresentation of environmental benefits and costs 17–18; financial performance pressure 19–20; interorganizational communication gap 20, 37; labor force 21, 37; lack of know-how and experts 15, 17, 37; middle management change inertia 20–21, 37; misconceptions about environmental issues 17; short-term profit effect on cleaner technologies adoption 18–19, 37
internal motivators and drivers 9–10, 26–30, *27*, 37; enhancing efficiency 30, 37; environmental accounting 29–30, 37; environmental leadership 28–29, 37; environmental management systems and continuous improvement 26–28, 37; environmental reporting 29, 37; voluntary initiatives 28, 33, 37
internal rate of return page??
International Organization for Standardization (ISO) 91, 150; 9000 28; 14000 26, 152; 14001 48, 49; 14021 93–94; 14024 91–93; 14025 93; 14040 141
international trade 7, 36, 37
Internet of Things 155
interorganizational communication gap 20, 37

just in time 12, 129

key environmental performance indicators 4

labor force 21, 37
lack of know-how and experts 15, 17, 37
Lancaster's theory 139
LCC *see* life cycle costing
lean supply chain 8
life cycle analysis (LCA) 79, 90, 91, 93–94, 161–168
life cycle assessment (LCA) 88, 141, 150–154, 161
life cycle costing (LCC) 44, 49, 60
life cycle of green procurement 44–50, *45*; application steps of set pair analysis 55–60 (*see also* application steps of set pair analysis; fifth stage: finalizing the contract 54; preparatory stage: choosing the green sourcing strategy 44–46; scoring

model **53**; second stage: determining the technical/environmental specifications 46; sixth stage: measuring the contract performance 54–60; third stage: screening of the green market 47–48; fourth stage: choosing green suppliers 48–54
life cycle sustainability dashboard (LCSD) 142–143
life cycle thinking 65, 72, 73

material intensity 71; per service unit 140
MCA *see* multi-criteria analysis
middle management change inertia 20–21, 37
misconceptions about environmental issues 17
multi-criteria analysis (MCA) 55, 111, 137, 139, 140

net present value (NPV) 138
Nordic Swan 92

obscured profitability 18
online analytical processing (OLAP) 159
outbound logistics (green transportation and distribution) 74–86; allocation of energy consumption *75*; allocation of GHG emissions 76; definition of green transportation 74–77; green practices in transportation/distribution 77–84
overview stage 67–68

paradoxes 11–13; cost 12; door to door 12; e-commerce sales channels 13; just in time 12; reliability 12–13; warehousing support 13
peer networks 34, 35, 37
polluter pays principle 18, 85, 109
prevention is better than cure 18

quantitative assessment of sustainability indices (QASI) 142

radio frequency identification (RFID) 155–157, *156*
recycled goods 26, 37
recyclable materials 46, 48, 71, 114
recycled materials 75, 100, 107, 117, 122, 126
regulation fostering pollution prevention innovation 31–32, 37
"Reinventing Environmental Regulation" 31
request for information (RFI) 47, 52
request for proposals 47, 52
request for quote (RFQ) 48, 52
research and development 32, 33
reverse logistics 97–135; available assessment alternative 103–106; product's life cycle 98; waste electrical and electronic equipment 99–103 (*see also* waste electrical and electronic equipment); waste management alternative strategies *97*; waste management and environmental policy 97–99
RFI *see* request for information

selective deconstruction 125, 127, 130, 131–132, *133*, *134*
self-regulation 32, 37
short-term profit effect on cleaner technologies adoption 18–19, 37
Silo Mentality 20
small and medium enterprises (SMEs) 17, 24, 25, 34, 41
strategic environmental assessments (SEA) 137, 138
supply chain pressure 34–35, 37
sustainability preference 72
sustainability reporting guidelines (SRG) 141

technologies of marketing and supply chain management synergies 146–168; green information and communication technologies (GrICT) 146, 147, 148, 149, 160, 168 (*see also* green information and communication technologies); *see also* auto identifications/sensors technologies; business intelligence; ecoCycle; information and communication technologies; Internet of Things; life cycle assessment; radio frequency identification
third-party logistics 83
top-down 9 *9*, 20, 131
total cost of ownership 44
total quality management (TQM) 27–28
traditional to green supply chain 8–11

vendor-managed inventory 148
volatile organic compound 63n1, 77
voluntary initiatives (internal motivators and drivers) 28, 33, 37

waste electrical and electronic equipment (WEEE) 99–103, 104, 105–106, 135; disposal 109, 111; hazardous materials **103**; heat treatment 108; integrated management 109–115, *110*, 113; multi-criteria assessment 111–115; pre-treatment 108; recycling 106–107, 108; stages of material recovery *105*, *106*; typical composition *101*; waste streams 99–103, 114, 115

For Product Safety Concerns and Information please contact our EU
representative GPSR@taylorandfrancis.com
Taylor & Francis Verlag GmbH, Kaufingerstraße 24, 80331 München, Germany

www.ingramcontent.com/pod-product-compliance
Ingram Content Group UK Ltd.
Pitfield, Milton Keynes, MK11 3LW, UK
UKHW022031061126
477531UK00003B/167

* 9 7 8 1 1 3 8 6 4 4 6 0 1 *